D1060688

Worlds Beyond

With the advent of space travel, we have begun to explore the worlds beyond Earth. Dedicated planetary scientists have spent their careers studying and exploring these worlds by telescope and through space missions. In this book, ten of these scientists have been brought together to describe their favorite worlds, what they have discovered, and what drives them to explore. Each tells a personal story. Together, the ten essays in this book range across the breadth of the Solar System – from hellish Mercury to the snows of Pluto. The stories these pioneers tell range from telescopic to robotic exploration, from personal adventures in the Antarctic to painting planetary landscapes, from the joys of success to the frustrations of failure. *Worlds Beyond* is the third in an extraordinary series of books in which Alan Stern has brought together leading space scientists to describe their work. The first of these volumes, *Our Worlds*, was truly a first of its kind in revealing the inner motivations of planetary scientists. Next, in *Our Universe*, leading astronomers and cosmologists explored the vastness of the Universe itself. Now, with *Worlds Beyond*, we return to our home – the Solar System – to visit fascinating new worlds and to explore them through the eyes of a second group of legendary planetary scientists.

Worlds Beyond

The Thrill of
Planetary Exploration
as told by Leading Experts

Edited by
S. ALAN STERN

PUBLISHED BY THE PRESS SYNDICATE OF THE UNIVERSITY OF CAMBRIDGE
The Pitt Building, Trumpington Street, Cambridge, United Kingdom

CAMBRIDGE UNIVERSITY PRESS
The Edinburgh Building, Cambridge, CB2 2RU, UK
40 West 20th Street, New York, NY 10011-4211, USA
477 Williamstown Road, Port Melbourne, VIC 3207, Australia
Ruiz de Alarcón 13, 28014 Madrid, Spain
Dock House, The Waterfront, Cape Town 8001, South Africa

http://www.cambridge.org

First published 2002

Printed in the United Kingdom at the University Press, Cambridge

Typefaces Hollander 10/15pt and Vectora *System* QuarkXpress [TB]

A catalogue record for this book is available from the British Library

Library of Congress Cataloguing in Publication data

Worlds beyond : the thrill of planetary exploration, as told by experts / edited by S. Alan Stern. p. cm.
Includes bibliographical references and index.
ISBN 0-521-81299-2–ISBN 0-521-52001-0 (pbk.)
1. Outer space–Exploration. 2. Solar system. I. Stern, Alan, 1957–

QB501 . W667 2002
523.2–dc21 2002025623

ISBN 0 521 81299 2 hardback
ISBN 0 521 52001 0 paperback

For my uncle, aunt, and three wonderful cousins,
Alan, Jean, Terri, Alana, and Kim,
Family

Contents

Preface

Modern human civilization now stretches back almost 300 generations, to the earliest organized cities. For most of that time, each clutch of humans identified their settlement and its surrounds as their home. Less than 100 generations ago, information transmission and transportation technologies were capable enough for people to form nation-states, consisting of many cities and villages and consider them as a new kind of "home." But now, in just the last two generations – with the advent of space travel – many people have come to see their "home" as the whole Earth. This concept would have been essentially unthinkable to the ancients, for the world was too large for their technology to integrate a world, or even a nation state, into an accessible and cohesive community.

So too, it is hard for us, now, to think of our "home" as being something even larger than our planet. After all, we are still trapped, both physically and to a very great degree intellectually on our home planet. A century ago, Konstantine Tsiolkovsky described the Earth as the cradle of mankind, and a cradle it *still* is. Yet, a logical extension of human history aft is that our descendents will very likely consider the wider realm of the Solar System as much a home as we, the descendents of ancient, regional civilizations, consider the whole of planet Earth as home.

For a few hundred humans, those who are planetary scientists, the vision of the entirety of the Solar System as our home is already becoming a familiar and natural concept. And it is this concept, in part, that gave birth to my idea to bring together a few of the very best planetary scientists in the world to write about their favorite other worlds, and in doing so to give a little perspective on what makes both they, and their favorite places, tick. I wanted them to tell some stories of planetary exploration through their eyes – the eyes of scientists who culminate spacecraft, laboratory, and telescopic explorations by interpreting, with human warmth, the cold 1s and 0s in computers that reveal the Solar System. Of course, with nine planets and almost 100 known satellites, not to mention myriad asteroids and comets, there were many choices of locale to describe, and many fine scientists to choose from to provide personal descriptions. But how to choose these ambassadors of our profession?

As a guiding concept, I selected scientists rather than worlds. I chose individuals who had shown a deep, career-long emphasis on and passion for some specific place they had been attracted to. All are known for being either particularly good speakers, or writers, or both. And all are members of the second generation of planetary scientists, trained or inspired by first-generation mentors who had seen the birth of our field at the dawn of the space age. These are the intellectual "sons and daughters," so to speak, of legendary pioneers like Kuiper, Urey, Sagan, and Shoemaker.

In 1998, this process culminated with the publication of *Our Worlds*, a suite of essays by planetary scientists about their favorite places. The popularity of this volume led, in 2001, to the publication of a similar volume on cosmology and extragalactic astronomy, called *Our Universe*. Both of these volumes were published by Cambridge University Press. Now, here, we present a second volume of essays about planetary science.

As I selected essay authors, I asked each to tell a personal story involving his or her own career and motivations, and to describe some aspects of a favorite world that they had invested long years exploring, and to tell their story from the heart. What follows is a set of wonderful and diverse essays ranging across the breadth of planetary science. The stories in this book range from innermost Mercury to outermost Pluto, from Antarctic to telescopic to robotic exploration, and from research to painting to covering planetary science. You will learn about the cratered plains of Mercury, the formation of the Moon, the atmosphere of Saturn's cold but strangely Earthlike moon Titan, and more.

The essay authors of this new volume have done a wonderful job. In this book you will find both a lot of planetary science and insider perspectives about how planetary science is done. So too, you will also see a good deal of what drives and interests planetary scientists, and, on occasion, you will see their inner hopes and aspirations revealed.

So, come and visit a few more worlds across the larger home that humankind is coming to know. Come and see a little bit of the heart within our science, and the hearts of planetary scientists. Come and visit, *Worlds Beyond*.

Alan Stern
November 2001

Introduction
Welcome home

Alan Stern

At night, when the last rays of the Sun vanish over the western horizon, and the brilliant blue is gone from the sky, a little something wonderful happens. We planetary landlubbers can suddenly see across the vast cosmic ocean to other shores.

And what a view it is! Across the cold but crystal clear expanse of vacuum so simply named, *space*, shines the light of a hundred billion stars within our home galaxy. A few thousand are close enough, and bright enough, to see with unaided eyes as individual lighthouses – truly these are beacons in the deep. So too, the dark night sky reveals a few galaxies, the nearest other star islands in God's Cosmic Pacifica. But it is not the galaxies that we seek, nor even the much closer stars of the Milky Way; what calls, most strongly, to soulful Earthlings, are the worlds that share the space around Sol, our very own Sun.

The ancients discovered the planets because they, unlike the fixed stars, moved across the night sky, changing their positions noticeably as the weeks and months slid by. And when the long eyes of the first telescopes were pointed heavenward, just four short centuries ago, they revealed that the five long-familiar wanderers of the night sky, "the planets," were wholly unlike the stars – for they appeared as rounded worlds, replete with clouds, polar caps, and changing seasons.

During the first ten of the twelve generations that have walked the green hills of Earth since telescopes revealed the planets as worlds, progress in understanding these places was slow. What the telescope could do well, when combined with a diligent astronomer, was to chart the Solar System and catalog its population. Telescopes revealed that most worlds (Mercury and Venus being the exceptions) carried with them about the Sun smaller worlds, moons, some of which were merely mountain sized, but others were large enough to be worlds unto themselves. With telescopes, astronomers also discovered that the Solar System was sprinkled with pockets of debris from the days of its origin: the asteroid belts between Jupiter's orbit and Mars, the Kuiper disk lying beyond Neptune, and the far away Oort Cloud.

But the telescope and the human eye alone were too feeble a tool to reveal much about the worlds they could see. Looking at worlds through a telescope is the astronomical equivalent of trying to perform a complete medical diagnosis with a stethoscope. Of course, that did not prevent the practitioners of astronomy from trying to picture the worlds that join Earth in orbit about the Sun – when you have a single tool, milk it! But the work went slowly and the results were meager. So over the first two and a half centuries of telescopic exploration of the Solar System, astronomers barely learned more than how to measure a few basic attributes of each world, such as its size, its mass, its temperature, and the length of its day.

As you might expect, these first facts no more revealed the richness and wonder of the planets than could the Mona Lisa be described by saying, "image of a young woman, 120 by 70 centimeters in scale, oils on canvas." The telescope, studying worlds from afar, reveals too little.

One thing that was missing was the ability to build accurate and sensitive cameras, spectrometers, and similar devices to make a useful harvest of the light that the telescopes gathered. And so, slowly at first, but then at an exponentially increasing rate, astronomers and engineers invented tools to dissect and record the light that shone down from each world to reveal something about its composition, its temperature, and its atmospheric makeup. But even with exquisite and increasingly sophisticated devices at the "business ends" of their telescopes, even with observatories perched on mountain tops high above the worst of weather and atmospheric turbulence, the planets were still too far away to see in much detail.

What was missing was the ability to set off from Earth and travel to the other worlds; to see these places up close; to come to know worlds by going to them; to map them carefully; to land on their surfaces, and probe their atmospheres; to touch and feel them; to bring home samples; to make them real. And when these steps became possible, our field left pure astronomy behind to become a new kind of science, a hybrid, somewhere between astrophysics and geophysics, called planetary science.

Cradle vista

Whoever it was that said that travel is broadening certainly had it right. The single most humbling lesson we have learned in planetary exploration is just how incorrect (and usually naive) our astronomically based perceptions of the planets have usually been.

Why was this the case? For one thing, telescopic observations generally produce too low a resolution to really see the details of the worlds we study. So too, telescopic observations are fundamentally limited in their ability to reveal the compositional and physical details that are possible with *in situ* measurements, such as wind speed, mass spectroscopy, gravitational harmonics, and seismic studies at a planet.

In essence, our ability to understand the planets from Earth was about equivalent to trying to fully understand the geology, climate, and cultures of, say, Asia, by flying over in a space Shuttle with a pair of binoculars. As a result, most of our *conceptions* about the planets, prior to the age of space exploration, were, well, *misconceptions*.

Remember? Mercury had an atmosphere, with clouds. Venus might be an inhabited swampland. Mars seemed to green with vegetation each spring. And the asteroids were thought to be the remnants of an exploded planet. The giant planets were uninteresting, except for that funny red spot on Jupiter, and only Saturn had rings. The moons of the giant planets were boring, and pretty much all the same.

None of these ideas, which were based on the best available evidence at the dawn of the space age just over 40 years ago, was correct.

To learn what the planets were truly like we would have to go there, but that was not easy. The first impediment was distance. Even with our best rockets, crossing the great, vacuous gulfs between the planets takes months within the compact inner Solar System, and years in the planetary outback beyond Mars and the main asteroid belt. The second impediment was (and to a large extent still is) the heavy engineering required to make spacecraft and their launch vehicles work reliably. Space pioneer Werner von Braun once said that, "We can beat gravity, but the paperwork is enormous." The third impediment was money, for the tools of technology are expensive. Even today, individual spacecraft often cost more than a small fleet of jet transports, and their launch vehicles usually about double that expense. Tom Wolfe was right, "No bucks, no Buck Rogers."

Nevertheless, the political necessities of the cold war forged a pathway to the planets in the form of a very public competition between the United States and Russia's Soviet Union. What were the two superpowers trying to prove? That each had a society that could lead the world. And lead they did, for the historic explorations they financed and executed will stand, as long as humans record their history, as positive testaments to the United States and the Soviet Union, and the curiosity, prowess, and ingenuity of twentieth-century civilization.

Free bird

So, owing to politics (as opposed to manifest destiny, or greed, or even scientific curiosity), beginning in 1959, we as a species went exploring other worlds.

The first steps were simply aimed at the Moon, less than 1% the distance to the nearest planet, for it was a time for learning the basics of how to launch and fly deep-space craft. And learn we did, as launch vehicles exploded, rockets went off course, spacecraft spun out of control, lost radio contact, or failed for one of a dozen other reasons. The Soviet records are sketchy, but US records show that, of the first four Pioneer and six Ranger missions to the Moon in the late 1950s and early 1960s, *none* fully succeeded, and only three of these ten missions could be called even a partial scientific success.

The first coup in planetary exploration came early, in 1959, when the Soviets launched Luna III on a week-long mission to obtain the first images of the heretofore-hidden far side of the Moon. Luna III's main scientific result was in discovering that the lunar far side has a vastly different appearance from the front side. This, like many early discoveries that were to come, was a lesson primarily about how little we had known, or could know, from studying far away worlds solely from Earth.

By the early 1960s, both the US and the Soviet Union had undertaken vigorous programs to make flybys of the two nearest planets, Venus and Mars. Only about half of the attempts succeeded. But when they did, they spoke volumes.

Later in the 1960s, while the robots made further forays to Earth's two nearest planetary neighbors, Apollo spacecraft delivered nine human crews to the vicinity of the Moon. Six of these crews were sent to the surface to deploy instruments, collect samples, and explore the geology of lunar mountains, valleys, and plains. It is a shame, but human exploration paused there, never to be restarted in the twentieth century and perhaps not even in the twenty first.

The 1970s saw Venus and Mars exploration move into a more sophisticated phase, with entry probes, landers, and globe-circling orbiters designed to make far more in-depth studies than simple flyby visits ever could. Later in the 1970s, the United States branched further afield, launching a triple flyby mission to Mercury and four different flyby spacecraft to Jupiter.

Few US missions were launched in the 1980s. However, the Russians continued their spectacular string of successes with the Venus exploration, and three of the US probes that reached Jupiter in the 1970s went on to reconnoiter Saturn. One of these probes, Voyager 2, was even sent on to make historic first explorations of the Uranus and Neptune system. Meanwhile, a flotilla of European, Russian, and Japanese flyby missions was launched at Halley's comet, and a US spacecraft was redirected to flyby another comet, called Giacobini-Zinner. By the time the 1980s ended, all of the planets save Pluto had been visited, and the Galileo mission was en route to make the first close reconnaissance of an asteroid, and then on to orbit Jupiter.

And what of recent times? The 1990s witnessed a re-flowering of planetary exploration, with more launches than in any decade since the 1960s.

Many of these missions were smaller and more focused than the expensive, "do-all" exploration missions conceived in the 1970s. It is a good thing too, because the larger missions had grown in complexity and cost to a point that they were not financially sustainable.

Today, as this book is being completed, we have just witnessed the highly successful, low-cost Pathfinder Mars landing, and the NEAR mission has recently completed the first detailed reconnaissance of an asteroid. So too, missions are now either underway or being built to explore Pluto, orbit and land on comets, return samples from asteroids and comets, put long-duration rovers on Mars, explore the multifaceted Saturnian system, and return to Jupiter's Europa to determine how far under its icy shell the water ocean lies. Planetary exploration is alive!

21st century

As we stand at the crossroads of time, the juncture of two centuries, and indeed, two millennia, we also stand at a kind of crossroads in the history of our species, and of our planet. Mother Earth has produced tens of millions of species, one of which, now, is taking its first tentative steps away from its birthplace. As we *homo sapiens* follow that course, in part, at least, for reasons we ourselves do not fully understand, we are coming to see the larger venue of the Solar System as an identity associated with our own. We are, slowly – but surely – gaining a planetary perspective beyond the cradle of Earth.

The Solar System is a vast place, but perhaps no vaster for us now than the Earth was to the ancients. We planetary scientists have enjoyed the pleasure and the privilege of being on the vanguard of the first, primitive wave of exploration across the sea of space and to the other shores warmed by the Sun. Come and see a little of what it is like to participate in this exploration, as we invite you into our home. *Welcome home.*

1

Dateline: the planets

J. Kelly Beatty, *Sky & Telescope* Magazine

Kelly Beatty has been writing about planetary science and astronomy for 28 years, and he is widely regarded as the best insider planetary sciences journalist in the business. I first met Kelly in 1987, just as I was preparing to enter graduate school, and then, as always, he impressed me as a scientist in journalist's clothing. Kelly grew up in Madera, California. He holds a Bachelor's degree in geology from the California Institute of Technology and a Master's degree in science journalism from Boston University. In 1986 he was chosen one of the 100 semifinalists for NASA's Journalist in Space program. Kelly and his wife, Cheryl, live outside Boston.

Tucked away in a corner of my office is an unmarked but bulging manila envelope. Inside, in no particular order, are nametags issued to me over a quarter century of scientific meetings, press conferences, and space-exploration spectaculars. I am not much for collecting souvenirs, mind you, but when I joined the editorial staff of *Sky & Telescope* (*S&T*) magazine in 1974 saving those mementos just seemed like the right thing to do. After all, there I was, at the tender age of 22, jetting off to legendary space places like the Jet Propulsion Laboratory (JPL) in California and Cape Canaveral in Florida. I was hobnobbing with famous scientists and high-flying astronauts, telling the world about their exploits, and getting paid for the privilege. What a job!

I did not set out in life to be a science writer – few of us in this journalistic niche do. No, I wanted to be an astronomer. Back in the early 1960s the space age was young, and like so many of my generation I was determined to be a part of it. I grew up on the outskirts of a small town in central California, under the kind of pitch-black skies that are now sadly so rare. No one yet knew what the Moon or Mars really looked like, and while skygazing through my backyard telescope I was free to imagine those landscapes untainted by what spacecraft images would later reveal. That zeal carried me through high school and on to Caltech, where the road to the mountaintop got pretty bumpy.

Becoming an astronomer was going to require more prowess with calculus and physics than I could muster, so I opted instead for a degree in geology. Besides, I rationalized, astronomers were more interested in stars than the Solar System – and I was in love with planets.

My first in-the-flesh exposure to science writers came in 1971, just about the time I started working for planetary scientist Bruce Murray. A specialist in Martian geology, Murray had been chosen by NASA to help analyze images of the red planet taken by Mariner 9. One minor role of that assignment called for him to participate in press conferences at JPL's von Karman Auditorium, and I often trailed along out of curiosity. The space-chasing press corps was an interesting mix of a few reporters who had a good grasp of things astronomical and a great many who did not. "I could do this," I remember telling myself, and thus was my fate cast.

The first nametag in my collection dates from 1976, but my first junket for *S&T* actually took place a couple years earlier. I had only been on the staff a few months when Mariner 10 made its first flyby of Mercury in September 1974. The pictures radioed to Earth showed a cratered world that looked outwardly Moonlike but which, on closer inspection, bristled with geologic enigmas. Murray headed Mariner 10's camera team, and at his suggestion I flew out to JPL to pick through the returned images to find some good ones for use in the magazine. So there I was, seated at a long table brimming with eight-by-ten-inch glossies fresh from Mariner 10's camera. I headed back to Massachusetts with a dozen or so that formed the basis of a "special report" for *S&T*. Meanwhile, the chosen images quietly ended up among those released to the general press. NASA's "fairness police" would never allow a lone reporter this kind of privileged access today, but it definitely impressed my new bosses.

Early on I learned that planetary scientists are very approachable. This was a great relief for someone trying to grasp the nuances of atmospheric dynamics, magnetospheric physics, and orbital mechanics. No question was too naïve or absurd to ask. These folks really wanted me to understand the concepts, to appreciate the subtleties, and to get the science right. Scientific endeavors rarely provide pat, black-and-white answers, nor do they lend themselves to quick sound bites. More often, they are works in progress that add incrementally to our collective knowledge, small brush strokes often painted in uncertain shades of gray.

Of course, exploring the Solar System via spacecraft often splashes the scientific canvas with a rush of pure discovery and high drama. Consider 1976s Viking mission, for example. In what ranks as NASA's most gutsy planetary mission to date, two spacecraft reconnoitered Mars from orbit, and two landers plopped down on to its surface. A team of geologists picked the locale for each touchdown, based on its inherent scientific interest and apparent smoothness. But in the end the chosen sites were only best guesses, and, as soon became obvious, the safe arrival of each lander proved a minor miracle. Pictures radioed to Earth revealed landscapes littered with huge boulders, any of which would have been a mission-ending obstacle had the automated descent sequences culminated a little bit this way or that.

Mars did not have the "magnificent desolation" of astronaut Buzz Aldrin's Moon. Instead, as seen through electronic eyes that painstakingly recorded their surroundings one thin scan line at a time, we discovered a world that seemed strangely familiar. In its Sun-drenched rocks and windswept ochre sands, we all recognized some familiar patch of desert in the American Southwest, a spot where we had scurried around rattlesnakes and clawed at the Earth. In fact, initially the Martian scenery seemed just a little *too* familiar: the first color image from Viking 1's lander showed a hazy, blue-hued sky. But a revised color adjustment revealed the sky to be more salmon than sapphire – a revision that drew hisses and boos from the press corps assembled in Von Karman. "Typical Earth chauvinist response," quipped Carl Sagan in reply.

The overriding reason for landing on Mars was to search for life, and once the suite of miniaturized incubators aboard each lander failed to find clear evidence of microbial inhabitants, public interest waned. They seemed to grow weary of seeing the planet's towering volcanoes, mammoth floodplains, deep canyons, and delicately layered polar caps. A news-media buzz followed the discovery of a hauntingly face-shaped mesa, but as the months dragged on it almost seemed that the mission had lasted too long. Only a handful of journalists camped out at JPL for more than a few weeks, and I was not one of them. In fact, due to *S&T's* spartan travel budget back then, I never made it to Pasadena at all. So I covered the mission *in absentia,* hoping the day's mail would bring a fat envelope from JPL stuffed with images around which to frame my next story. And those stories continued for an unexpectedly long time: Viking 1's lander kept sending data until November 1982, more than six years after touching down. By then my attention had shifted outward from Mars to more distant worlds in our Solar System, and to a pair of intrepid spacecraft that were making astounding discoveries about them.

Voyager and "instant science"

If Viking qualifies as NASA's riskiest planetary mission ever, then Voyager must surely rank as its greatest planetary-exploration adventure. Launched in 1977, the mission's twin spacecraft were equipped to take a close look at Jupiter, Saturn, and whatever else whey encountered on their long and winding road out of the Solar System. The first of these meetings played out in March 1979, as Voyager 1 swooped close to Jupiter and its quartet of large, "Galilean" moons. The flyby would reveal much about the planet, of course, but the geologist in me was itching for the close-ups of Callisto, Ganymede, Europa, and Io. Discovered in 1610 independently by Galileo Galilei and Simon Marius, two of these satellites are the size of our own Moon and two the size of Mercury. It is no stretch to think of them as "worlds." Yet despite centuries of telescopic observation (and some cursory scrutiny by Pioneers 10 and 11), we knew very little about them. The Voyager flybys would change all that.

Voyager 1's rendezvous with Jupiter was a bona fide media event, and JPL literally teemed with hundreds of reporters. There were big-time television personalities, newspaper veterans, space cadets who had snuck in as stringers for their college newspapers, local radio and TV reporters, and a galaxy of foreign correspondents from Europe, Japan, and Australia. Production trucks lined the street leading up to von Karman Auditorium, their roofs bristling with microwave antennas and myriad cables snaking into them. In his book *Distant*

Figure 1.1

Geologist Larry Soderblom has a captive audience of reporters as he describes Voyager 1's images of Rhea and Dione, two of Saturn's satellites, in November 1980. He and other mission scientists had expected to see little more than impact cratering, but they were hard pressed to explain the wide range of unusual terrains found on the moons' surfaces. (JPL image P-25139).

Encounters, writer Mark Washburn dubbed this spectacle the "attack of the space gypsies."

Each day the horde would assemble for a press conference, at which project scientists would dispense pictures and other data beamed to Earth only a day or so earlier. This show-and-tell approach was unfamiliar territory for the mission's researchers. Accustomed to spending months, even years, wringing results from their observations, they were being asked by NASA to leap into the rarefied air of "instant science," with no safety net and, literally, all the world watching. Moreover, when it came to the most newsworthy results, there was a fine line between releasing exciting, front-page images and holding something back for later publication in professional journals. After all, many on the team had devoted the better part of a decade to the Voyager mission, and to give all their hard-won results away in a flurry of press conferences just did not seem fair.

Not everyone was comfortable with these arrangements. It did not help that television monitors throughout the pressroom showed us Jupiter's swirling clouds, the enigmatic cracks on Europa, and other eye-widening vistas as soon as they were received. We gypsies speculated freely among ourselves about what the images showed. Moreover, this was not the largely uninformed press corps of the Mariner era. Spurred by new popular magazines like *Discover*, *Omni*, and *Science 80*, a new generation of reporters with solid scientific credentials was filling the chairs in Von Karman. The questions were more pointed, and our collective appetite for technical detail more voracious, than JPL's media-relations team had ever seen.

The engagement never became adversarial, however, due in no small part to the mission's spectacular discoveries. The first flyby revealed the Great Red Spot, Jupiter's signature cloud feature, to be a cyclonic maelstrom with a voracious appetite for smaller clouds trying to skirt along its periphery. Io surprised almost everyone with the ferocity of its volcanic activity, and Callisto bore an enormous impact whose concentric rings looked like expanding ripples frozen in place. Voyager 2, which called on the planet four months later, found even more eruptive activity on Io and revealed Europa to have an exterior that bore more than a passing resemblance to a heavily cracked eggshell.

By the time everyone returned for Voyager 1's sweep past Saturn in November 1980, the scene at JPL had the air of a college reunion. Scientists and reporters were mingling freely, renewing acquaintances, and consequently the press conference patter became more genial and less guarded. Once again, there was plenty of dazzle to go around. The rings of Saturn defied all predictions, resolving under Voyager's scrutiny into thousands of individual ringlets that left everybody at a loss for answers. When confronted by the discovery of mysterious dusky "spokes" in the rings, Bradford Smith, the mission's imaging-team leader admitted, "I don't think there is anything that has kept us puzzled for so long." Theorist Jim Pollack quipped, "We're still pretty much in the gee-whiz phase." And so it went.

Despite some infirmities, Voyager 2 reached Uranus in 1986 and Neptune in 1989, thus completing a "grand tour" of our Solar System's four largest planets. Dynamicists first realized in the 1960s that a

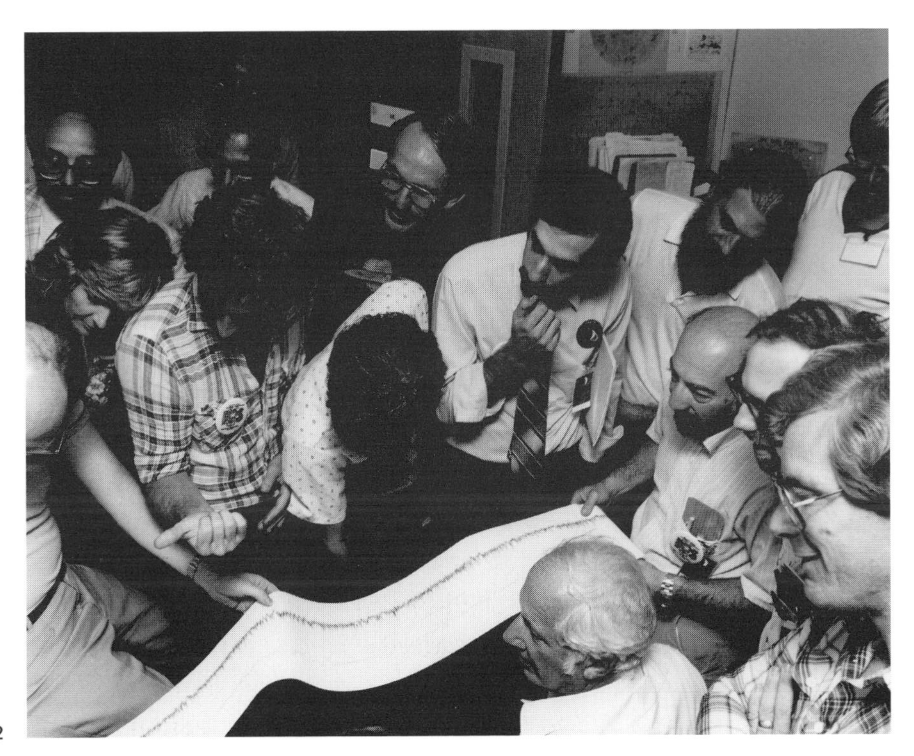

Figure 1.2

An exciting moment for scientist A. Lonne Lane (at end of computer printout) during Voyager 2's flyby of Saturn in August 1981. His photopolarimeter (a sensitive light-measuring device) had malfunctioned on Voyager 1, but the one on Voyager 2 performed superbly. The printout shows how a star's light varied as it passed behind the thousands of individual strands that make up Saturn's rings. The author stands behind Lane at extreme upper right. (JPL image 24022).

spacecraft could carom across the outer Solar System in just 12 years, using a boost from one world's gravity to speed to the next. The planetary alignment needed for these slingshots would occur in the late 1970s – and not be repeated for another 176 years – so it was an opportunity not to be missed. Mission designers had equipped the Voyagers as best they could for a four-planet trek, even though Uranus and Neptune were always technical and political long shots (in fact, they were not official mission objectives when the spacecraft left Earth). Thus we were all grateful that Voyager 2 survived long enough to reach them. Upgrades to NASA's receiving network here on the ground greatly enhanced the aging spacecraft's scientific return, and the hits just kept on coming. Uranus proved as remarkable for its climatic blandness as did Neptune for its storminess. Triton, a large moon circling Neptune in a retrograde ("backward") direction, revealed itself to be geologically young and still smoldering with activity.

There is a little-known footnote to this celebrated mission. By design, Voyager 1's planetary objectives would end with Saturn, whereas its twin would try to press on to Uranus and Neptune. Their respective trajectories were such that Voyager 2 had to be launched first, and its Titan-Centaur rocket performed flawlessly on August 20, 1977. But during the launch of Voyager 1 two weeks later, on September 5, the Titan booster shut down prematurely. The Centaur upper stage barely compensated for the shortfall, reaching Earth's escape velocity a scant three seconds before exhausting its fuel. Voyager 1 barely had enough oomph to reach Jupiter. By pure chance, Voyager 2 got the better rocket – had it been paired with the underachieving Titan instead, its encounters with both Uranus and Neptune would have been lost.

Culture shock

Although NASA has been in the thick of space exploration since its creation in 1958, Americans have occasionally had to stand aside while other nations' efforts took the spotlight. Such was the case with the return of Halley's comet in 1986, when the United States opted not to send a spacecraft to intercept this infrequent but celebrated visitor. It is commonly thought that NASA passed on this opportunity for lack of money, but the real reason was a lack of desire. In 1981 the planetary exploration committee of the National Academy of Sciences' Space Science Board ranked such a mission "of markedly lower priority" than other planetary projects awaiting approval. Neither NASA nor the newly installed Reagan administration chose to overrule that position, despite an intense, vocal lobbying effort mounted by Bruce Murray, who by then had become director of the Jet Propulsion Laboratory.

Part of the Space Science Board's reluctance to endorse JPL's Halley Intercept Mission was that other nations were already well along in building comet chasers of their own. Soviet scientists took a couple of their standard-issue deep-space chassis, added cameras and other instruments, and formulated a plan that would swing the craft past Venus en route to the comet. This duo came to be called Vega, a contraction of the Russian words Venera ("Venus") and Gallei ("Halley"). Meanwhile, the European Space Agency's directors were pressing ahead with a spacecraft called Giotto, whose compact drum shape and special shielding would help it survive a plunge deep into the comet's dusty coma and close to its icy, hidden nucleus. Even the Japanese joined the Halley armada, with two craft – Suisei ("Comet") and Sakigake ("Pioneer") – that would take measurements from afar.

For an American press corps that had grown fat on NASA's near-perfect string of success stories (*Challenger* had not yet been lost), the prospect of covering missions whose control centers lay across the Atlantic and Pacific oceans – not to mention on the far side of the Iron Curtain – was a little daunting. Among other things, in 1982 the Reagan administration had allowed a scientific-exchange agreement between the US and USSR to expire, so contact with Soviet space officials had all but ceased.

But, as it turned out, my curiosity about the "Venus" portion of the Vega mission paid an unexpected dividend. At that time the Soviet Union's civilian space science program was controlled by Roald Z. Sagdeev, a plasma physicist with near-fluent English and numerous colleagues in the West. I contacted Sagdeev (by telex!) through one of these intermediaries, asking whether I might come to Moscow to cover the Vegas' arrival at Venus in June 1985. OK, he replied. But given America's strained relations with the "evil empire," getting to Moscow would still prove complicated. First, I needed an official invitation from the Soviet Academy of Sciences, essential for obtaining an entry visa. Only after getting the visa could I then arrange for transportation and lodging. Somehow it all worked out, and on the morning of June 11 I took a seat in the main auditorium of the academy's Space Research Institute, known widely by its Russian acronym IKI. Among the dozens of assembled scientists I was the only American – *and* the only reporter.

By then, 100 million kilometers away, a spherical canister dropped off by Vega was plunging into the hot, murky atmosphere of Venus. Some 60 kilometers up, a small parachute opened, and the top half of the probe cleaved away as its heavier bottom continued in the dark toward the planet's hellish surface. Dangling from a parachute, the discarded cap disgorged a small package that quickly inflated into a balloon some three meters across. In a few minutes this buoyant bubble was floating free, its French-built instruments dangling below from a 12-meter-long tether.

Back in Moscow, the *mood* of the event – more than the language – was foreign to me. I had become accustomed to the spirited hubbub that accompanied major space events at NASA centers. But at IKI that day there was hardly any conversation, let alone hubbub. A large projection screen displayed flight parameters like elapsed time, the balloon's altitude, and the temperature of Venus's sulfurous atmosphere. When word came that data from the instrumented balloon had reached Earth, as heralded by an anonymous voice booming out of some speakers, no one applauded. In fact, the scientists continued their passive vigil right through the announcement that the large descent probe had landed safely on Venus's surface. From atmospheric entry to touchdown, the first Vega encounter lasted 65 minutes. Then the auditorium emptied.

Vega 1's triumph made the front page of *Pravda*, and four days later the sequence of events played out again for Vega 2. With one success already assured, this time the mood was more celebratory. The French ambassador was on hand, as were a handful of Soviet reporters. At a press conference later that day, team members described a few early results, including results from the two balloons. But no one mentioned (as I learned later) that the first lander's drill had failed to obtain a surface sample for analysis, or that two experiments on Vega 2 had malfunctioned.

I did not realize it at the time, but my modest involvement at IKI that June would carry over to the following March, when the Vegas made their long-awaited dashes through Halley's dusty coma. This time Sagdeev allowed reporters from several American publications to attend, though he asked *me* to select them on his behalf – an "honor" that I politely declined. Once our troupe arrived in Moscow, I was further embarrassed to learn that the Academy had appointed me, together with Rick Gore of *National Geographic* (who had also previously visited IKI), to head the journalist "delegation." For that I endured months of good-natured teasing from my colleagues.

Both Vega spacecraft survived their 79-kilometer-per-second dash through Halley's coma, the mission scientists were more relaxed about revealing what had worked and what had not, and we all got our stories. Then our little entourage quickly left town for the small German town of Darmstadt, 25 kilometers south of Frankfurt, from which the European Space Agency (ESA) would guide its Giotto spacecraft to Halley's nucleus.

A veritable circus greeted us there. The Halley encounter still stands as the grandest space adventure (and largest media event) in ESA's history. Television trailers beamed news of the mission to 50 nations and an estimated billion viewers. Giotto carried ten experiments, most to study the coma and its interaction with the solar wind, but in the eyes of the world there was only one: its camera. The first images, obtained early on March 14, about three hours prior to the moment of closest approach, bore little resemblance to telescopic views of the comet. That is because the camera team had chosen to display them in a garish rainbow of computer-generated colors, perhaps to captivate the worldwide television audience. But, if that had been the hope, it backfired, instead causing great confusion as to how the nucleus really looked.

Giotto pressed on, slicing through the coma at nearly 250,000 kilometers per hour. Then, just nine seconds before zipping to within 600 kilometers of the nucleus, the screens went blank. A grain of dust, perhaps as small as 0.1 gram, had whacked the spacecraft and caused it to wobble. Battered but not beaten, Giotto righted itself and reestablished its radio link with Earth a half hour later. (It undoubtedly heard the applause still echoing through the quiet streets of Darmstadt.) The best of its 2,000 images certified that Giotto had come face to face with the heart of Halley, a hulking iceberg some 15 kilometers long and eight kilometers wide whose surface was as black as coal. Giotto had confirmed what telescopic observers had already suspected: Halley was large by cometary standards, far surpassing the one-to-two kilometer diameter commonly found among its siblings.

The success of Giotto catapulted ESA to the status of major player among space powers, but the agency also learned that its press relations needed work. It did not help that the individual experiment teams, rather than ESA itself, controlled the release of their results.

I remember walking the control center's deserted corridors on March 16, a Sunday, just one day after the sole post-encounter press conference. I was looking for someone – *anyone* – to interview, but by then most everyone had already left town. A rare holdover was Harold Reitsema of Ball Aerospace, one of just two Americans on the camera team. Sensing my desperation, he handed me the most precious keepsakes of my entire two-week odyssey: three unambiguous, unadulterated, unbelievable black-and-white images of the comet's nucleus. It was a sympathetic gesture I have never forgotten.

In the years since Giotto, ESA continued to develop wonderful and highly productive space science missions. And today the rainbow graphics are a distant and amusing memory. ESA's public relations office has morphed into a well-honed, media-savvy information machine. Ironically, NASA's efforts have trended in just the opposite direction: its current space-science missions are *required* (not just encouraged) to develop their own plans for public outreach. The result, unfortunately, is a decentralized mishmash of Web sites where the PR wheel is constantly being reinvented.

The "Great Crash"

Astronomical discoveries are usually made far from our purview, at some isolated control center or from a remote mountaintop observatory or by a spacecraft dashing somewhere in the Solar System. Rarely are we afforded front-row seats to these discoveries – let alone to some event that changes the course of science. But that is exactly what happened in July 1994, when the fragments of a shattered comet named Shoemaker-Levy 9 slammed into the planet Jupiter. For once, anyone with a decent backyard telescope could witness history in the making.

This cosmic saga began in March 1993, when the observing team of Eugene Shoemaker, his wife Carolyn, and long-time collaborator David Levy, along with visiting French astronomer Philippe Bendjoya, settled in for a night of asteroid hunting using a modest research telescope at Palomar Observatory in California. But soon thickening clouds forced them to call it a night. Two days later, while examining what little film they had shot, Carolyn found a fuzzy streak not far from Jupiter. "I don't know *what* this is," she said, bolting upright. "It looks like . . . like a squashed comet." Larger telescopes later revealed the streak to be a line of little comets arrayed like pearls on a string, each sporting its own tail.

In time, astronomers realized that a single object had strayed too near Jupiter and been torn apart by the planet's gravity – and that the comet's remains were destined to strike Jupiter itself. No one could predict the outcome with certainty, though a cruel twist of geometric fate had placed the target zone on Jupiter's far side, just out of sight from Earth. Computer-aided simulations tried to anticipate what would happen during each high-speed splash into Jupiter's atmosphere. Some modelers assumed that the fragments were large, at least a mile across, and would strike with the kinetic-energy equivalent of 100 billion tons of TNT or more. Others thought Jupiter might swallow the shards without a trace.

The "Great Crash" played out over six days, beginning July 16, as a score of large fragments bombarded the planet at 40 miles per second. Never before had such an event been witnessed, and never before had so many of the world's telescopes turned their gaze to the same spot of sky. Yet the news media focused the lion's share of attention on the Space Telescope Science Institute in Baltimore, Maryland, which serves as the clearing-house for observations from the orbiting Hubble Space Telescope. It was there, during a news conference on the evening of July 16, that television cameras captured one of the most exuberant and spontaneous displays of scientific joy ever recorded.

As the comet's discoverers offered cautious predictions for what would be seen when the first chunk of comet smashed into Jupiter's atmosphere, observer Heidi Hammel rushed in with a near-infrared image fresh from Hubble's camera. A huge dark feature stained Jupiter's southern hemisphere, marking a titanic splash of impact debris. Hammel uncorked a bottle of champagne and exchanged high-fives with the Shoemakers and David Levy, as the assembled press corps roared its approval.

In the days that followed, several impacts created fireballs thousands of miles high, tall enough to peek around the planet's limb and be spotted by the Hubble telescope. As each of these atmospheric wounds rotated into full view, still fresh and hot, they looked like titanic flares when recorded by infrared detectors on ground-based telescopes. Most of the crash zones lingered as huge, dark stains in the Jovian atmosphere (some larger than the entire Earth) that took weeks to fade away.

In hindsight, the "Great Crash" proved to be a watershed event for astronomers. Eager to share their results with their colleagues worldwide, and keenly aware of public interest in the event, many observers threw their telescopic images onto Internet web sites almost as fast as they were taking them. They realized that the tedious analysis of these data would consume many months, if not years. Writing up their results, and waiting for scientific journals to accept and publish them, would take even longer. But for once the peer review and the nit-picking conference debates could wait – it was enough to let the dramatic pictures speak for themselves.

This was not the first time that scientists had exploited the power of the Internet to sidestep the traditional routes for distributing results, and certainly it would not be the last. In an era when good publicity can boost one's chances of getting funding, astronomers and other researchers continually wrestle with how best to announce their findings. The urge to "publish" electronically is strong, and professional journals are struggling to maintain their relevancy. Some have already moved wholesale into the Internet arena; eventually I suspect they all will do so.

Comet Shoemaker-Levy 9 may also represent the last big science story for which news coverage followed traditional paths. That is, people learned about it by and large from periodicals and broadcasts – the Internet had not quite emerged as a news machine. *Sky & Telescope* was right in the thick of it too, but, because it is published monthly, we could not provide up-to-the-minute crash reports. Instead, we adopted an editorial plan that would provide our readers with a "behind-the-scenes" perspective.

So I hit the road, heading first to Palomar Mountain in California, then to Mauna Kea in Hawaii, to watch teams of observers in action. Visiting Palomar was, in itself, an exciting event. Enraptured as a boy by my first view of the legendary 200-inch telescope, I was thrilled to stand beneath its massive frame at last and to touch the housing for its hallowed mirror. It was a religious experience. Hawaii, by contrast, was a letdown: the weather had turned nasty, and none of the mountaintop telescopes could open their domes to observe. Eventually, I retreated to my hotel room, pounding out a story while other staffers back in our editorial office scoured the Internet for the latest images.

We moved Heaven and Earth to get nine pages of coverage into our readers' hands four weeks later. All things considered, it was a worthwhile effort. Yet had the Great Crash occurred just two years later, in 1996 instead of 1994, our approach would have been very different: most of our resources would have been diverted to getting the story on to our Web site as soon as possible. Filling the magazine's pages would still have been important, of course, but we would likely have taken a slower, more measured "what-does-it-all-mean" approach.

The road less traveled

Such heady, dramatic events are the exception, not the rule. Most of the nametags in my collection have been gathered at rather ordinary meetings of professional researchers – one day spent at a workshop in Providence; a weeklong affair in Tucson; and everything in between. I really thrive on these, mostly, I suppose, because it is one small way that I get to participate as the scientist I have always wanted to be. In that environment I work on my own, without the safety net afforded by press conferences and spin doctors. Instead, I must bring to bear all my technical and journalistic savvy in search of newsworthy results.

The key (for me, at least) is to become attuned to the lingo of research. Sometimes the title of a presentation is enough to pique my interest. One recent example was "Meteor Storm Evidence Against the Recent Formation of Lunar Crater Giordano Bruno," presented as a simple poster by a University of Arizona graduate student named Paul Withers. There has long been speculation that in the year 1178 medieval monks witnessed the collision of a small asteroid into the Moon, a wallop that purportedly blasted out a 22-kilometer-wide crater. But Withers had deduced that the whole idea was preposterous, because such an event would have showered the Earth with ten million tons of fragments – creating perhaps a trillion bright meteors – in the days thereafter. And no mention of such an awesome display appears in English, European, Arabic, or Asian chronicles of the era. It made for a very satisfying story: a little history, a little observational astronomy, and the can't-miss appeal of a catastrophic impact (even if it never occurred).

Sometimes researchers cloak newsworthy results in obscure terminology. Consider, for example, this prosaic talk title from a recent meeting: "Spectral Feature Mapping with Mars Global Surveyor Thermal Emission Spectra: Mineralogic Implications." Geologists Roger Clark and Todd Hoefen had used spacecraft data to compile a map of minerals exposed on the Martian surface, and they had found widespread evidence for a particular mineral called olivine. Olivine is rather common on Earth, but it weathers away easily in the presence of water. So the upshot is this: if olivine exists all over the Red Planet, water could never have been widespread on its surface. You would never guess that from the title, nor even from the authors' published summary. Now, in fairness, not everyone believes that those spacecraft maps are being interpreted correctly. But if they are, the implications for the climatic history of Mars are profound.

Results aside, one of the great joys of attending professional meetings is getting to meet scores of graduate students and post-docs. Much like the medical profession, it takes many years to turn someone into a proper "scientist." But planetary work has the advantage in that eager, up-and-coming students can often make lasting – even significant – contributions to our knowledge. So I make a point to get to know as many of them as I can. And, having now done this for a couple decades, I have watched a generation of them grow up. Some become "leading experts" in their field; others leave the profession behind, for their families' sakes, for greener pastures, or sometimes out of discouragement.

If covering space exploration for a quarter century has taught me anything, it is this: despite the stereotypes about astronomers forced upon us by television and movies, most planetary scientists are not geeks. In fact, quite the opposite is true. They are generally a lot more well rounded and talented than my everyday friends. I know Solar System specialists who are competitive runners, who perform in opera companies or rock bands, and who are accomplished cooks. They get married, have children, and move from job to job, just as we all do. These very human qualities are rarely seen or appreciated by the public at large.

In fact, the public gets precious little exposure to the real work of science at all. For most of us, first-hand experience with chemistry or biology or astronomy ends well before high-school graduation. After that, our scientific awareness is largely in the hands of the news media, whose representatives pass on to us those salient topics considered timely and that have enough news value to compete with the daily potpourri of other headlines. Before these stories can reach you, in whatever medium, those of us who create them must convince our higher-ups that the stuff is worth publicizing.

No question about it, Mars landings, Voyager flybys, and comet crashes rise to an interest level high enough for even grizzled newspaper editors to take notice. When a spacecraft discovers ancient rivers on Mars, or a probe makes a risky dash through the dust-choked coma of a comet, the news value is so obvious that the stories practically write themselves. And if it is a NASA mission, a small army of media-relations personnel will be on hand to make sure reporters are handed stacks of fact sheets, or get pictures to publish, or arrange for interviews with key personnel. It is almost like walking in with a bucket and having the news generously ladled into it. ("One scoop or two, Mr. Beatty?")

The problem is that fewer and fewer space-exploration stories qualify as "big." For starters, these days most interplanetary missions are headed for places we have already visited at least once. The rush of pure discovery has largely passed. Moreover, we live at a time when various distractions are vying for the public's scientific attention. Today astrology is as popular as ever, pseudo-science has been legitimized by television, and "government conspiracies" lurk around every corner. Compared to all that, real astronomy can sometimes seem downright boring.

Example: on September 22, 2001, an aging, crippled spacecraft called Deep Space 1 dashed through the coma of a big comet called Borrelly at more than 16 kilometers per second. Moving at that speed, a mote of dust packs the punch of a bowling ball. Against the odds, Deep

Space 1 survived, and it radioed to Earth incredible pictures of the comet's coal-black nucleus. I wrote four different stories about this encounter in one week for *S&T* and its associated Web site. Some newspapers saw fit to cover the story in depth. Yet the *Boston Globe,* a major daily newspaper, published just a single picture and a short caption, with no hint of the science and drama that was inherent in this risky mission. The *Globe*'s readers deserved better. Granted, we were all still gripped by the horrifying terrorist attacks in New York and Washington that had occurred only 11 days beforehand.

These issues of science and journalism beg two big-picture questions: Is the exploration of our Solar System still worth doing, and is it still newsworthy?

To the first I would offer an unqualified "yes." We are far from knowing everything we would like to about our neighbor worlds. Kids still get jazzed about seeing Saturn's rings through a telescope, and scientists still get jazzed when the discussion turns to how those rings came to be and how they "work." (The Cassini orbiter should offer some important insights when it arrives there in 2004.) No matter how comprehensive or voluminous, each mission's results always prompt a new spectrum of provocative scientific questions.

However, like it or not, funding for science comes mostly from governmental agencies – whether in the US or elsewhere – that are subject to the political posturings of a given leader or national mood. When President John F. Kennedy exploited the Apollo program to earn technological bragging rights, funding for NASA and the National Science Foundation surged, and legions of Baby Boomers began dreaming of careers as space scientists and engineers. But when government leaders show indifference or set their priorities elsewhere, the pace of space exploration slows. Even scientifically provocative missions are deferred or canceled. Doing so does not render the rationale for our scientific quest invalid; it just means that we must sometimes be patient.

To the second question, I offer a more tentative "yes." Science and journalism often exist in a love–hate relationship. Researchers bemoan shallow, sensationalized stories, yet increasingly the news "buzz" created by a particular research project or space mission can play a role in setting its funding priorities – the heightened interest in astrobiology being one recent example. Unfortunately, in an attempt to get exposure for their work, scientists (or their institutions) sometimes tout a finding as newsworthy when it really is not. NASA's public relations machine alone churns out hundreds of press releases annually, many more than can be accommodated even in a special-interest publication like *Sky & Telescope.* This shotgun approach, in itself, is nothing new; reporters have always been faced with sorting out what constitutes "news."

What has changed, however, is our ability to access not only the news stories but all the primary source material itself – press releases, press conference videos, whole scientific papers, and even raw spacecraft data are now available via the Internet. Those of us who hunger for the latest astronomical findings no longer need newspapers and the evening news as the delivery media, so we no longer clamor as loudly for their inclusion by those sources. Taken to the extreme, everyday reporting about astronomy and other sciences might someday disappear from the mass media altogether.

I, for one, do not want that to happen. We explore space for the benefit of all, not just the informed few. And as long as the majestic, mysterious universe continues to enthrall and inspire us, I am confident that we will find value in vicariously exploring its uncharted depths and savoring what we discover.

Further readings

Hartz, Jim and Rick Chappell (1998). *Worlds Apart: How the Distance Between Science and Journalism Threatens America's Future,* Nashville: First Amendment Center, Freedom Forum.

Sagan, Carl (1995). *The Demon-Haunted World: Science As a Candle in the Dark,* New York: Random House.

Washburn, Mark (1983). *Distant Encounters: The Exploration of Jupiter and Saturn,* New York: Harcourt Brace Jovanovich.

2

Planetary science with a paintbrush in hand

William K. Hartmann, The Planetary Science Institute

I first met Bill Hartmann in the late 1980s when I interviewed for a position at the Planetary Science Institute in Tucson. At that time, Bill was already legend in the field of planetary science, an accomplished author, and one of the foremost space artists in the world. Bill, an expert in the study of Mars, the Moon, and the asteroids, was the first winner of the Carl Sagan Medal of the American Astronomical Society, and was recently elected a Fellow of the American Association for the Advancement of Science. He was an originator of the currently accepted theory of the origin of the Moon, co-discoverer of multi-ring impact basins on the moon, and an imaging team member in NASA's Mariner 9 and Mars Global Surveyor missions that laid the groundwork for understanding Martian geology. His most recent novel is called, *Mars Underground.*

When I was a kid in the 1940s, my older brother had an encyclopedia called the *Book of Knowledge*, and in the M volume was an article on the Moon that contained a map. The map enthralled me. It had names! It is one thing to know there are worlds up there in the sky but it is something else to know they have places on them: mountains, plains, and valleys with names!

A few years later, probably in 1950 when I turned 11, I got a book that actually showed those places as they would appear to a human visitor. It was *The Conquest of Space* with text by German rocket pioneer Willy Ley and paintings by a man named Chesley Bonestell. The paintings showed scenes throughout the Solar System: an eclipse of the Sun by Earth, as seen from the Moon; Saturn's rings as seen from a vantage point above Saturn's clouds; dusty plains on Mars; Saturn hanging in a blue sky over the icy plains of Titan (the blue sky being based on Gerard P. Kuiper's discovery of Titan's atmosphere in 1944). Bonestell became a boyhood hero. Later I realized a dream, meeting him when he was in his eighties, and conversing with him about painting.

Bonestell, as described in a 2001 biography and retrospective by Ron Miller and Frederick Durant, was trained in architecture and paid special attention to realism and geometry, calculating the angular size of Saturn seen from Titan, Jupiter from Ganymede, and so on, and fitting the planet, with its calculated angular size, into a picture of specified angular width (which, like the choice of telephoto or wide angle lenses, controls the appearance of shadows and perspective of foreground features such as craters). For example, if the planet is 20 degrees across, and the painted image represents a typical snapshot angular size of 40 degrees, then the planet must be half the width of the painting.

The idea that you could study the planets not just as abstract physical objects of investigation, but as places, and figure out what it would actually be like to be there, shaped my view of science. The scientific goal for me was not just a career making measurements or deriving mathematical models, but also a personal quest to find out what the rest of the universe was about. What would it be like to stand on Mars on fly into the Orion nebula? Did other stars have planets? If so, what would they look like? And did they have people on them? What was humanity's relation to the rest of the universe? Even as a teenager, I had a half-subliminal idea that you could not have a real philosophy of life unless you had some valid picture of the rest of the universe. And these seemed like questions whose answers could be expressed by exciting images as well as by exciting equations. Images and equations are just different ways of knowing. To go out further on a limb, it seemed to me in some indefinable way that if you could not paint a picture of another planet or star system, you really could not claim to understand its nature. I relate this to the history of natural science. In the 1500s, art and science were two branches of the same thing: understanding how humans relate to the universe. Leonardo da Vinci practiced art to be a better scientist, and science to be a better artist. He sketched and painted nature in order to understand how things work, and at the same time he dissected cadavers to understand anatomy so he could paint better portraits. It was all a mix, a continuum.

The Copernican revolution of the 1500s and 1600s allowed humans for the first time to see planets not as supernatural entities in the sky, but as real places, putting humans not in the center of the picture, but rendering them only as part of the grand scheme. It is a lesson that we humans still have not fully learned. Newspapers still carry astrology columns, tabloids and TV producers purvey junk science, and state legislators pass bills to ban the teaching of evolution or replace it with ideological propaganda. Scientists have somehow not communicated and disseminated the exciting view of the universe that has been painstakingly cobbled together in the last few hundred years. The Copernican revolution is continuing, and I suggest that teamwork between scientists and artists will play a role in advancing a better view of our real place in the cosmos.

Art and science: mixed roots

My grandfather, Andrew Hartmann, was a painter. He was classically trained as a painter and craftsman in Switzerland in the late 1800s, then came to the US and set up a studio in St. Louis (see color plate section, Figure 2.1). He apparently tried to be a fine artist, but this was America, and we all know what that means about making a living as an artist; so, after he was married, he supported his family as a house painter and decorator. Still, some sort of creativity bubbled to the surface and in his seventies and eighties, when I knew him, he was still producing landscape paintings from Pennsylvania to California. Many of these were unsold, or sold to friends at laughably low prices. My other grandfather, Robert Carmichael, grew up on a farm in post-Civil War Alabama, educated himself, and ended up as head of the Mathematics Department at the University of Illinois. He published some poetry, and, according to family lore, he published some of the earliest American papers on relativity. As I grew up, there was a sense that Grandad Hartmann was "only a humble painter," while Grandad Carmichael was an exalted college professor. However, now, as I look back, I see them as equal in creativity, and I think their relative social status involves some deep issues in the American paradigm.

As Grandad Hartmann's paintings appeared on the walls of our house while I grew up, I did not realize that my own future would trace an arc that would go from a Ph.D. in astronomical research, to the painting of landscapes on Earth and on other planets. I often wonder what all this says about the ghostly role of inheritance. Is there some brain wiring involved? In the intervening generation, my Mom taught high-school math before settling down to raise a family, and Dad was an engineer who ultimately headed a research lab developing and applying alloys for the Aluminum Company of America. I think some of his practical approach rubbed off on my research style. And, much as I try to experiment, my painting style comes out looking like my Grandad's. Perhaps I was subliminally influenced by his aesthetic. And what about my other Grandad's poetry? I have not gone the poetry route, but I have published textbooks, popular science books, and a novel about Mars, and another novel about the Southwest is under contract. The interesting thing is that in none of this science, painting, or writing, did I ever have any conscious drive to emulate my grandparents. It just seemed to come out, and now I look at it and wonder. . . In any case, painting and writing seem more and more to me a natural evolution from a career in science, and a way to heighten my own understanding of the relation of humans to nature. You take in a certain amount of experience about "life, the universe, and everything," to use Douglas Adams' phrase, and then it is time to process it and put it back out in some new form.

Big pictures

I want to talk about the odd connection between science and painting, and how it seems to teach a lesson – that you have to back off and look at the relationship of the big facts, rather than getting lost in the details. I did not fully realize that this was my view of science until much later, in my forties and fifties. At that time, people began to tell me that I was a "big picture guy" or that I seemed to think visually, though I did not think that last statement was true, since I had published theoretical as well as observational material. So I began to think more about the enterprise of science itself, and how it was done, and its relation to art. By looking at my own career as a subject of investigation, I discovered not only that I probably was a "big picture guy," but also that this stemmed from my view of the universe. It may have to do with functions of the brain's hemispheres. I am not as good a left-brain (stereotyped?) analytic/mathematical scientist as a lot of my colleagues, but I think there is a need in science for the right-brain (also stereotyped?) spatial/relationship ways of handling information as well as left-brain linear sequencing way of handling data. I was pushed into thinking about this sort of thing because my colleague, Carl Sagan, once gave me a copy of his novel, *Contact*, with an inscription about my "functioning corpus calossum." I rushed to the dictionary to see if Carl had insulted me and learned that the corpus calossum is the nerve bundle connecting the two hemispheres! You do not get a comment like that from someone like Sagan without starting to think about your own work! Armed with that comment, I began to analyze the different kinds of scientific thinking.

There is very strong pressure during a science career to become a narrower and narrower person. You write a Ph.D. dissertation. Your colleagues demand that you publish parts of it, present it at meetings, and become the world expert on that little piece of the universe. Then someone comes along and publishes a different theory. You are now expected to respond, to be the spokesperson for this subject. Every time someone else publishes a paper on that subject, you have to read it and then rush to the web or the library to delve ever deeper into your subject. Pretty soon you are the reigning expert on the size of boulders in ejecta blankets thrown out of impact craters, or Xenon isotopes in the stratosphere or Neptune, or bee's wings or dinosaurs' toes, or winds blowing over the polar caps of Mars, or some other tiny facet of nature. The pressure is toward ever-finer levels of detail. All of scientific history is commonly viewed as a march toward ever-finer levels of detail.

Einstein reportedly said that "God is in the details," and when I give talks I like to shock audiences by claiming Einstein was wrong! For the sake of argument, I like to propound that God is in fact not in the details but in the first-order information, the big picture. It is more important to know that pi is approximately 3 than to know that the fifth significant figure of pi is 6. In spite of the way that accountants make us balance check-books, the number of dollars you have is a hundred times as important as the number of pennies. So, in human life, our basic perception of reality depends more on the first-order facts than on the third-order details.

To insist that God is only in the details is like studying the pixels in a digital image one at a time, and literally missing the big picture.

Although this is an important principle in ordinary life, it is not a popular view in science. Organized science likes detail just as organized religion likes ritual. When a scientific paper is published, it must pass a referee process. This means the paper is sent to colleagues, who review it for accuracy and plausibility. I claim that the usual referee report boils down to "The author needs to discuss the data in more detail (and, by the way, the paper needs to be shorter)." In my own papers, I have often been forced by referees and editors to go back and add more discussion of trivial nuances in the data, even though they have not the slightest effect on the final conclusion of the paper. "That's the way science must be done!" they say. And that is true, at some level. Yet the big advances in science come when someone reaches a new understanding not of detail but of grand, underlying relationships, as when Einstein and others in his generation overturned Newton by seeing new ways to treat space and time. The differences between Einstein and Newton, in ordinary circumstances, appear only in the tiniest details, but the revolution came from a new look at the most fundamental ideas and assumptions.

These issues are at the root of why scientists have so much trouble communicating their work to the public. The onlooker wants the overview, not the details.

I would claim that a painter learns to understand these ideas instinctively. As an example, I can quote a Russian artist friend. During the Glasnost period of Soviet history, the International Association of Astronomical Artists was invited to the Soviet Union for some space art exhibitions, and I had the opportunity work with a number of Russian artists as well as scientists. Because so much of Soviet culture froze in 1917, they seemed to have a direct pipeline back to the techniques and ideals of the impressionists and other master painters at the beginning of the twentieth century. I had the opportunity to paint in the Moscow studio of Andrei Sokolov, the leading painter of the Soviet space program. He allowed me to make a small canvas that would be taken up to the first great space station, Mir. As I worked excitedly with ever-smaller brushes, trying to paint in details, he looked on and frowned.

"What's wrong," I asked.

"Bill," he said, "when I was a student, my master taught me, perfect is the enemy of good."

What he meant was what every painter eventually learns. I have several versions of this same story. I was painting outdoors one day with my friend, the well-known southwest landscape painter Pete Nisbet. Pete had studied with the great master of southwest landscapes, Wilson Hurley. "Tell me some lesson you learned from Hurley," I asked Pete. "It was a basic lesson," Pete said. "The lesson was "back off!"

What the lesson meant was that you need to back away and look at the whole picture, not get lost in the parts. You can ruin a good painting by trying to fill in too much detail. The thing that makes a painting yield some kind of truth about a scene is the right balance of the first-order elements: overall composition, broad tones, balance of colors, brightness, darkness, and so on. In a sense, this is the other end of the scale from classical photography, which records every detail, regardless of importance, in a kind of left-brain linear logic. Mastery of this concept led the impressionists away from the rendering of every detail, and to focus more on the qualities of light itself, for which they were lampooned as painters of mere "impressions." Yet the impressionists, contrary to popular image, were not dreamy romantics but scientists of light. They were very involved in studying the technical literature of the 1800s about how light interacts with objects and eyes. Their attempt to render colors by combinations of pure primary tones was inspired by scientific findings about the spectrum.

Using my own career as a guinea pig, I can give several examples of what I mean by saying that first-order reality is as important as the details. The first example came around 1961–1962 when I was a graduate student working under Professor Gerard Kuiper in Arizona. President Kennedy had announced that we were going to the Moon, and the fledgling planetary science community was scrambling to understand lunar geology. Naturally, owing to the preoccupation with detail, there was a huge push to take ever-sharper photos of the Moon. The best photos showed features about one kilometer across, and everyone tried to push to the next level of detail. Maybe some features 700 meters across would reveal the great secret of the Moon! But the

Figure 2.4

On the Moon. This lunar landscape attempts to capture the effect of low light, at sunrise or sunset, among the lunar rocks. Apollo astronauts generally landed at midday in some of the smoother, safer lunar regions, and thus brought back photos only of some of the less-spectacular lunar views. Many more vivid effects of geology and lighting remain to be experienced by future explorers! Painting by William K. Hartmann.

biggest advance, in my personal experience, came from the other direction. Kuiper had set up a system to project the best lunar photographs on a globe, and then re-photograph the globe from the side, to show the lunar surface without foreshortening. Remember that the Moon keeps one side toward Earth, so that humans normally have only a foreshortened view of features around the edge of the Moon, and we never see the back side. As soon as we projected photos on the globe, we could walk around to the side of the globe and see the relationships between lunar geographic details that no one had seen before. Within the first months, we realized we were seeing giant concentric ring structures that had never been adequately recognized. The 1890s-era American geologist, G. K. Gilbert, the 1940s crater expert, Ralph Baldwin, and the 1950s-era cosmochemist, Harold Urey, had described certain parts of these systems around Mare Imbrium, but no one had realized that there was a class of giant impact features, sharing the same characteristic multi-ring circular basins, with concentric rings spaced apart by a factor of 2, and surrounded by radial lineaments, all with point symmetry around the basin center.

The best example was the Orientale basin system, which jumped out as an astonishing bulls-eye of rings on the extreme eastern edge of the Moon. I went to Kuiper with these ideas and in 1962 we published the discovery paper on the Orientale basin and other general classes of concentric basin features. At the time, this was simply an exciting discovery, but in retrospect it strikes me as amazing that we could make a fundamental advance by backing off from the Moon to find new 1,000-kilometer-scale patterns, when everyone else was trying to push to the next level of detail and study 700-meter scale features!

A second example comes from our work at the Planetary Science Institute in Tucson (known as "PSI") on what has emerged as the currently accepted theory of the origin of the Moon. After the discovery of the giant lunar basins, I had been studying impacts on the early planets by giant asteroid-like bodies, left over from the origin of the planets. The multi-ring basins on the Moon, like Orientale and Imbrium, needed impactors about 100–150 kilometers across. I began to wonder: what were the largest bodies that hit the early planets. Meanwhile, Victor Safronov, in the Soviet Union, had been publishing papers about the collisional aggregation of the planets, and the likelihood that Uranus had been hit by a giant body to tilt its rotation axis to the observed tilt of roughly 90 degrees (as opposed to Earth's familiar 23.5 degrees, which causes the seasons). I had read some of Safronov's papers in an obscure journal of Russian translations, *Soviet Astronomy*, but they were little known in the west. At the same time my colleague Donald R. Davis, at PSI, was making theoretical calculations of how fast planetary bodies could grow from colliding asteroid-sized bodies. So I proposed we work on the following problem: as Earth grew to its present size, how big might the second largest bodies have grown in that same zone of the Solar System? As we worked on this, we concluded, somewhat to our surprise, that the second largest body could have reached the size of the Moon or even of Mars by the time all early planet-forming material was used up – if it was lucky enough to avoid hitting Earth (or getting thrown out into some other planet after a

Figure 2.5

Mir and Shuttle: a lost opportunity. For a period in the 1980s and 1990s, the Russian and American space program had beautiful complementarity. Russia had a space station and big boosters, America had the shuttle and good miniaturization. The total mix offered a fantastic opportunity for humans to expand their capability in space, but it was largely ignored. The American shuttle did perform a number of docking missions with the Mir station, but no major joint planetary missions were designed. When planetary scientist Robert A. Brown and I tried to lead a move toward joint utilization of the Russian/American capabilities while serving on the National Academy of Science "Committee on Planetary Exploration," the discussion was cut off and we were told that political support could not be found in Washington to support such a program. Painting by William K. Hartmann.

close encounter with Earth) at a still earlier stage. If such a body reached that size and then hit the growing Earth, it could blast out material that might have re-aggregated into the Moon.

I realized this could explain the puzzling lack of iron in the Moon. Even before Apollo, everyone recognized that the Moon has the same low density as rocks, rather than the higher bulk density of Earth with its huge iron core. This meant that the Moon has no substantial iron core, and this gross, first-order fact had frustrated most earlier theories of the origin of the Moon. But, if an early giant impact blew debris into orbit after Earth's iron core had formed, the ejected material would be iron-poor, rocky material from Earth's upper mantle – just the kind of material from which the Moon was formed, as proven by the lunar samples brought back by Apollo astronauts and Russian Luna probes.

In 1974 I gave our paper at a conference on satellites. After my talk, Al Cameron, a revered figure in the field, stood up to say that he and his younger colleague, Bill Ward, were independently working toward the same idea, but starting from angular momentum considerations, instead of from growth and rock chemistry considerations. They were reaching a similar conclusion and thought the giant impactor had to be as big as Mars! We published our paper in 1975 and they published an abstract of their work in 1976. Most of our colleagues in the 1970s were loathe to accept catastrophic impacts, and the seemingly wild idea languished until a 1984 conference, when various speakers evaluated various lines of evidence. By the end of that conference week, our idea emerged as the leading theory of lunar origin. The idea has been developed much further by workers such as Robin Canup, who tells her own story later in this book.

What does this have to do with details versus first-order facts? Again, I realized this connection only in retrospect. I had actually begun to ask people about my giant impact idea at lunar science conferences in the early 1970s. Redoubtable geochemists assured me that it would not work because they had found certain third-order isotopic differences between terrestrial rocks and lunar rocks. I was in awe of these geochemists, because they could measure chemical compositions and ages of rocks to four, five, or more significant figures, whereas Don Davis and I could only calculate crude first-order growth rates. I actually held back for a year or two before working on these ideas, because my geochemist friends assured me the Moon could not be made of ejected Earth material! Yet the advance came by thinking more about the first-order properties (the mean densities, the iron cores, the likelihood of occasional big impactors) than the third-order details. As our impact idea began to be accepted, the chemical details began to be explained by geological processes involving magma evolution.

A third example of the importance of first-order thinking comes from work that Dale Cruikshank, Dave Tholen, and I did in the 1980s on comets and asteroids. This was a case of looking at the first-order nature of the phenomenon rather than getting hung up on the details of semantics. The semantic distinction between comets and asteroids was Victorian, based on their appearance in nineteenth-century telescopes. Comets are fuzzy and asteroids are star-like. Comets give off gas, and asteroids are rocky. As a result of this deeply ingrained paradigm, comets and asteroids ended up being studied by two different groups of people. Comet scientists were gas spectroscopists, and asteroid scientists were geology types. Comets were discussed in certain sessions at certain meetings by certain people, and asteroids were discussed at other sessions by other people. In reality, knowledge in the 1970s suggested that a continuum of interplanetary bodies existed, from the rocky (ice-poor) "asteroids," in the inner Solar System, to ice-rich "comets," from the cold, outer Solar System. Bodies cataloged as asteroids had been found in the remote outer Solar System, but most were too faint for detailed spectroscopy. Our simple idea was to study the hypothetical relationship between the two classes of bodies by using one set of instruments, and looking not at spectral details, but rather comparing the overall colors. This could be done on much fainter, more remote bodies than were needed to study detailed spectra. To make a long story short, what we discovered is that these two categories ("comets" and "outer Solar System asteroids") had similar, overlapping ranges of color and spectral properties. In the process, we also made the discovery that the interplanetary object Chiron, which had been cataloged as "asteroid 2060," suddenly brightened up, blew off a cloud of dust, and turned into a comet, which proved that you could not trust the artificial semantic distinction that had been built into our thinking about such objects.

Comets were supposed to be "dirty icebergs," and so people had pictured them as white with a bit of entrained dust. But we found that Chiron, outer Solar System asteroids, and distant inactive comets were all very dark grey or brownish-black. Just before Halley's comet came by in 1986, a major review paper in the journal *Science* repeated the old wisdom that comets were likely bright, citing the most likely reflectivity as about 28%, a light grey color. In 1985, we had already published our conclusion that Halley was likely black, and would have only 4% reflectivity, probably due to sooty carbon compounds known to exist in the outer Solar System. This paper was roundly criticized, and, at a meeting in Flagstaff, Dale was blasted for the whole premise. The "linear logic people" argued that you cannot say anything about the reflective properties by measuring color.

Technically this was right, but they missed that we were synthesizing several known sources of first-order information and looking for patterns of similarity. A few months later, the European probe Giotto flew past Halley's comet and measured its reflectivity at 4%. Comets are icy, but it is very dirty, sooty, dark-colored ice. The criticism Dale received in Flagstaff had been so intense that a writer for *National Geographic* remembered the scene and later asked Dale, "After that scene, how does it feel to be proven right?" The point here is not whether the reflectivity was 4%, 5%, or 6%, but the basic underlying idea that comets and asteroids did not have to be divided into two distinct classes, and could be investigated by a scheme that backed off from detail and compared overall patterns of color similarity.

These examples all suggest to me that the basic way of looking at things, and the ability to back off and look from a distance, can be as important as the details discovered. This is true in science, art, and life in general. Science would not advance and cell phones would not work without the study of detail, but humanity also will not progress without people backing off and looking at "the big picture" – completing the revolution

In science, we break nature down into digestible units, studying specific aspects such as the chemical composition of rocks. In understanding nature through painting, the way Leonardo did, similar issues come up. In the first place, we have to reverse that process and synthesize everything we know about a place, how it came to be, and how light interacts with it under different conditions. The best astronomical paintings, from decade to decade, thus form an important record in human history – they synthesize what we human beings thought other places in the universe were like during that decade. We have to distill a sense of place as best we can in each painting, and each painting thus, at the same time, crystallizes a moment in astronomical history.

The Czech writer, Bohumil Hrabal, tells a funny story about a painter in his famous, dark novel, *Closely Watched Trains*. "I'm painting the sea," says this fellow. "I'm enlarging a seascape from a picture postcard."

"Why not come out and paint it from nature?" asks his friend.

"If I painted from nature, I'd have to reduce everything instead of enlarging it."

After I read this, I decided we painters and planetary scientists have a related problem. To us, Earth has become too small, a picture post-card. We need to enlarge, back off, get out and experience a larger reality. If all goes well, humanity as a whole will follow this path, and we can complete the Copernican revolution.

Further reading

Cameron, A. G. W. and William Ward (1976). The Origin of the Moon (abstract), in *Lunar Science VII*, Houston: Lunar Science Institute.

Cruikshank, Dale P., W. K. Hartmann, and David Tholen (1988). 2060 Chiron. International Astronomical Union, Circular No. 4653.

Cruikshank, Dale P., W. K. Hartmann, and D. J. Tholen (1985). Colour, Albedo, and Nucleus Size of Halley's Comet, *Nature*, **315**: 122–124.

Hartmann, William K. (1981). Discovery of Multi-Ring Basins: Gestalt Perception in Planetary Science, in Multi-ring Basins, Proceedings of the Lunar Planet Science Conference, 12A, 79–90.

Hartmann, W. K., D. P. Cruikshank, and J. Degewij (1982). Remote Comets and Related Bodies: VJHK Colorimetry and Surface Materials, *Icarus*, **52**: 377–408.

Hartmann, W. K., D. J. Tholen, and D. P. Cruikshank (1987). The Relationship of Active Comets, "Extinct" Comets, and Dark Asteroids, *Icarus*, **69**: 33–50.

Hartmann, W. K. and Donald R. Davis (1975). Satellite-Sized Planetesimals and Lunar Origin, *Icarus*, **24**: 504ff.

Hartmann, W. K. and G. P. Kuiper (1962). Concentric Structures Surrounding Lunar Basins, Communications Lunar Planet Laboratory 1: 51ff.

Hartmann, W. K., David J. Tholen, Karen J. Meech, and Dale P. Cruikshank (1990). Chiron: Colorimetry and Cometary Behavior, *Icarus*, **83**: 1–15.

Kuiper, G. P. (1944). Titan: A Satellite with an Atmosphere, *Astrophysics Journal* **100**: 378ff.

Miller, Ron and Frederick C. Durant III (2001). *The Art of Chesley Bonestell*, London: Paper Tiger.

3

Chasing new meteorites, or finding heaven on earth

Michael Zolensky, NASA Johnson Space Center

Mike Zolensky is one of the nicest people you would ever want to meet. He took his Ph.D. in geochemistry at Penn State, after taking a geology degree at the New Mexico Institute of Mining and Technology. His specialty is in the study of meteorites and space dust, most particularly, their chemistry and crystal structure. In addition to having published over 400 research papers, he is also involved in space missions like Stardust and MUSES-C. We met many years ago at a chili cook off at a Lunar and Planetary Science Conference, where he immediately impressed me with his quiet charm. Mike and his wife Sandra and their young daughter Jean live just outside Houston.

How do stars form, evolve and die? How and when did our Solar System form? What were the building blocks of planets, and why did they form as they did? Where did water come from, and how and when did it arrive at our home planet? What were the precursor organic molecules of life, and are these materials available anywhere else in the Solar System? How much extraterrestrial material hits the Earth every day, and has this bombardment been constant back through time? If something falls out of the sky and hits my mailbox, will I get rich? These fundamental questions can be addressed through the recovery and investigation of meteorites.

Meteorites have been at the heart of my career for almost 20 years – ever since I made the decision to exit the world of nuclear waste disposal, which was the topic of my graduate studies. In the course of that former work, at Penn State University, I had by necessity become very adept at handling and analyzing microscopic quantities of matter. It turns out that the most primitive meteorites have grain sizes so small most scientists become discouraged, but I had developed just the right skills to step into this new and exciting field and make useful contributions.

I am an employee of NASA, working in mosquito-ridden Houston, at the Johnson Space Center (JSC). JSC is known mainly as the home to the astronauts, and for spacecraft engineering. However, because of the necessity of training the Apollo astronauts in geology, a small lab devoted to the study of lunar rocks and meteorites was established here in the 1960s. The heart of this lab is the curation facility for lunar rocks. Later we added meteorites and interplanetary dust to the list of materials that we study and care for. While there are certain disadvantages to being a civil servant, I will not dwell on them. In fact I have something of a dream job: I can work on problems largely of my choosing, as long as they are important steps towards the ultimate goal of determining the origin and evolution of our Solar System, and making this knowledge available and understandable to the US taxpayers, who pay my salary and fund the research.

We know that most meteorites come to Earth from asteroids, but a few come from the Moon and Mars. There is reason to hope that some may even come from the planets Mercury and Venus, although the odds of this are slim and no such meteorites have been recognized. Mercury is located deep inside the Sun's intense gravity well, and it is very difficult for materials to creep out and travel to Earth. The atmosphere of Venus is 100 times denser than that of Earth, and 10,000 times that of Mars, and it is very difficult for impact-ejected rocks to punch through such a thick barrier and make it out to space. Still, with meteorites entering the Earth's atmosphere every day, who knows what may turn up?

Sometimes the meteorites come to you

Hardly a week goes by without my receiving an email or a phone call from some excited person, who has found something so out of the ordinary that they believe it just has to be a meteorite. Most people are satisfied to show me their find, and accept the usual verdict that it is just an ordinary rock, or something man-made. Some people are harder to convince. There was the young man who had found about a ton of fossil-laden rocks in a lake, had laboriously hauled them to dry land and installed them in a special hideaway in the woods. He was willing to part with them for only 100 million dollars. I kept the video he sent as a souvenir, and returned his rocks. One woman sent me a jar containing a small stone that she swore was giving off a strange liquid and moving around on its own power. I kept that one on my office window for about three months to make good and sure it was dead. It was. I returned it to her. Recently a man sent me a rock sealed into a jar with some leaves and fungus. According to him the rock was giving off a radiation that was undetectable with a Geiger counter. He knew this because the leaves inside of the jar were not decaying, and if one balanced a glass of ice water on top of the jar the ice would not melt. I returned the jar to him unopened, and told him that OSHA did not permit me to handle materials that gave off strange radiation undetectable with a Geiger counter. (This sounded like a real regulation to me.) Nearly everyone is excited by meteorites, and are tantalized by the possibility of actually finding one. All meteoriticists have similar stories to tell. We all spend a huge amount of time investigating these claims because every once in a blue moon, something truly wonderful does turn up this way.

Here is a story of one such amazing discovery. In the early evening of March 22, 1998, residents of an approximately 70-mile swath around the arid west-Texas town of Monahans reported sonic booms and an eerie, streaking light. The fireball and the accompanying fall of one stone were observed by young boys playing basketball outside – in fact the meteorite dropped to the ground only a few feet from the hoop. The boys picked it up within a few seconds and reported that it was still warm. A second stone was recovered the next day from an adjacent street by a local police deputy. Both stones were taken into police protective custody. The total recovered mass of both stones was 2.5 kilograms, about 6 pounds, although it is likely that other individuals from this fall escaped discovery. Working quickly, Everett Gibson, a meteorite scientist working at JSC who had spent his boyhood in a town nearby Monahans, flew out to meet with the boys and town officials. Everett is the sort of friendly, straight-talking Texan who can wheel and deal with the best, and he finagled the loan of both stones for preliminary characterization at our lab. Since there was already a meteorite found in Monahans in 1938 (some towns are especially lucky), the new meteorite was named Monahans 1998.

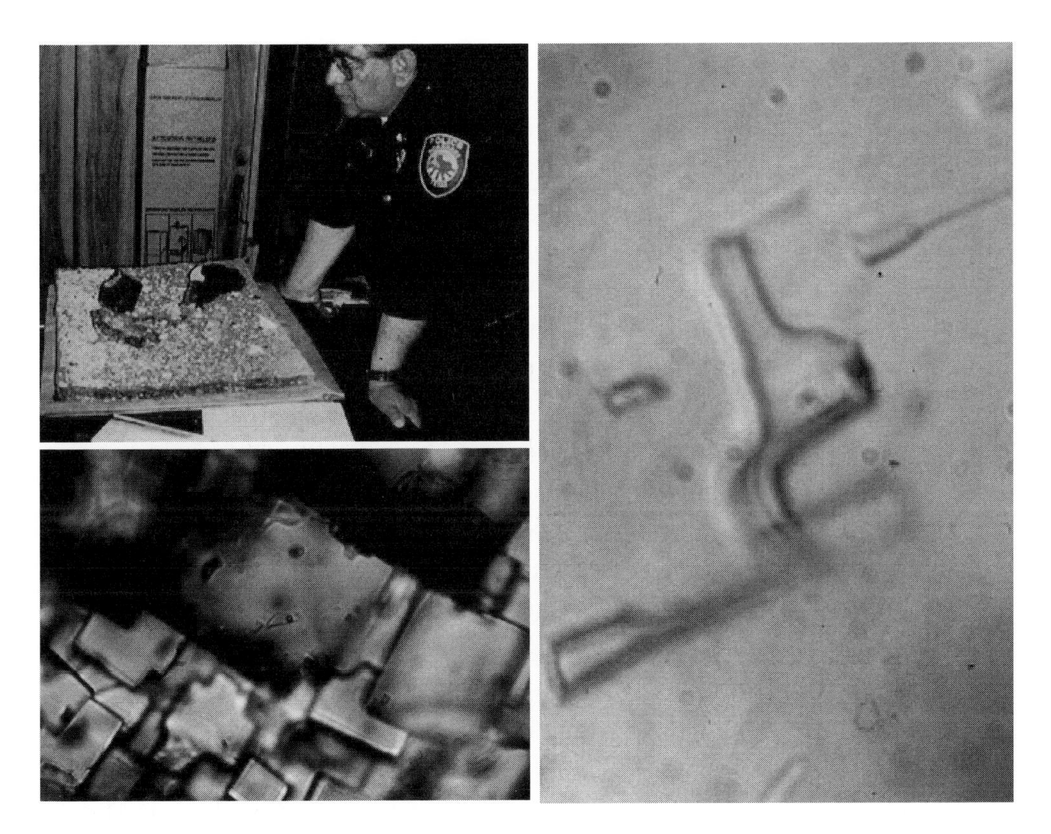

Figure 3.2 Monahans meteorite recovery. Upper left: Meteorites in police custody. Photo courtesy of Everett Gibson. Lower left: Crystals of halite, with tiny fluid inclusions visible inside. Right: Fluid inclusions at higher magnification.

The very next day, meteorite scientist Kathleen McBride broke open one of the Monahans stones in the spotlessly clean Meteorite Processing Laboratory at JSC. It proved to be the most ordinary of meteorites (an H5 ordinary chondrite). Kathleen also noticed that there were beautiful purple grains inside, something she had never before seen (and she looks carefully at thousands of meteorites every year). Kathleen called this to my attention and I set a small chip of the meteorite into our scanning electron microscope for a closer look. In addition to offering a greatly magnified view of a sample, this instrument is capable of determining chemical composition in a crude way. At the time I believed this to be the mineral sodalite, a rare but typical mineral in these meteorites, although the color was a bit off. Sodalite is navy blue, and the mystery crystals were dark purple. To my astonishment, the chemical analysis showed the mineral to be halite, NaCl – the same thing as table salt, with minor inclusions of the related mineral sylvite (KCl). I was so dubious of this amazing result that I even tasted a tiny chip, in the time-honored field mineralogist fashion. It was indeed salt! I began to wonder whether it had been such a good idea to taste it. Too late to worry about that now, but I am paying closer attention to my health these days.

Halite and sylvite had been previously reported as microscopic grains within two other meteorites, but nobody had believed the reports. The reason is that on Earth these minerals typically form from the evaporation of huge quantities of water, and are found in sedimentary rocks. Nobody has ever found a sedimentary rock as a meteorite. We reported the discovery of the Monahans halite at a couple of scientific conferences that year, and made the prediction that it might actually be a rather common mineral in meteorites. It's just that meteorites are typically rained on before recovery, or are washed, or sawed open using water or oil, and these actions would always dissolve any halite that might be present. Fortunately an alert meteorite dealer named Edwin Thompson (called "ET" by his associates) was paying close attention to our reports. Several months later he contacted me with the news that he had found what he believed to be halite in another new meteorite, called Zag, which had just fallen in Morocco. This halite was a beautiful sky blue color, but otherwise identical to that in Monahans. Scientists quickly provided three independent age dates of the Monahans and Zag halite, and all dates were the same – 4.5 billion years old. This halite dated from the very dawn of our Solar System.

It is well known that exposure of halite to ionizing radiation produces the same blue-to-purple color as is observed in Monahans and Zag, and we now believe that this coloration was produced in the meteorite either by exposure to solar and galactic cosmic rays, or by exposure to radioactive ^{40}K (in the sylvite) while in space. The essential point is that the blue to purple color proves that the halite has a pre-terrestrial origin, since simple terrestrial halite is colorless (just check your salt shaker). Several interesting possibilities are suggested by the halite occurrence in these meteorites, but the most likely explanation is that it formed from the evaporation of asteroidal salty water (called "brines"), in space!

When we thought about this situation for a while, we became very excited. When sea-water evaporates on Earth and crystallizes as halite, microscopic droplets of water are always trapped inside of the growing halite crystals. These are called aqueous fluid inclusions. If we could locate fluid inclusions in the meteoritic halite it would represent our first and only samples of water from asteroids. Now, many primitive meteorites contain abundant quantities of clay minerals, carbonates, sulfates, and other minerals that we are confident must have formed from water-rich regions of asteroids. We have known for decades that water was not only present but abundant for a time early in asteroid history. However, the origin and fate of this water has always been a total mystery. The resolution of this mystery has been the main focus of my career.

My next step was to call on my old friend, Bob Bodnar (Virginia Polytechnic Institute). As fellow graduate students at Penn State, Bob and I had shared an office, a passion for mineral collecting, and we worked together briefly on the characterization of fluid inclusions from ore deposits. While my career took a strange turn into meteorite research upon graduation, Bob had gone on to relative respectability as the world's leading authority on fluid inclusions. One afternoon, soon thereafter in my office, Bob and I set a few of the Monahans halite crystals on a glass slide, placed them under my petrographic microscope, and took a look. In less than one minute Bob turned to me and said, "there they are." I sat down and took at look at the first samples of water ever found from off the Earth.

Why are these samples so exciting? Well, consider that we still do not know where the Earth's water came from. If you look in planetary science books they will typically tell you that this water all came from comets. However, recently, scientists have learned that the ratio of the two most common isotopes of hydrogen (deuterium and hydrogen) in terrestrial water (including the water making up most of your body) are in a very different ratio than that observed in the coma of comets. What this indicates is that most of the Earth's water probably arrived from another source. Maybe it arrived inside grains of halite and other minerals in meteorites, or in larger bodies – actual asteroids that may have hit the Earth early in its history. Maybe the water droplets inside of the Monahans halite crystals are surviving remnants of the water that has made the Earth such a beautiful world, so hospitable to life. Now for the first time we have the chance to solve this fundamental mystery.

Searching for meteorites – not as difficult as you think, if you know where to go

Strategies for meteorite recovery are varied. You could stand in your back yard with binoculars and constantly scan the skies, and some very patient people try this. However, at the measured flux at the Earth's surface of one meteorite per square kilometer per 100,000 years, you have to be exceedingly patient and extraordinarily lucky. A more productive strategy has been to consider the forces that act to destroy fallen meteorites, and perform systematic searches of places on Earth where these forces act very slowly. Liquid water is the greatest enemy of meteorites, as it will inevitably dissolve even the toughest iron meteorite. In fact, some types of meteorites can be destroyed by a single gentle spring rain. Another sinister factor is erosion; streams will abrade and carry meteorites away to certain destruction. Therefore it is clear that dry environments are the prime targets for meteorites searches, and this means deserts. Fortunately for me, I happen to love exploring deserts. Now deserts are not all water free. In fact the most productive place for meteorite recovery is the continent of Antarctica. Here we find the largest reservoir of fresh water on Earth, but in a perpetually frozen state. Antarctica is the coldest, driest, highest, windiest continent, and all of these attributes serve to concentrate and preserve meteorites. However, none of this occurred to anyone, and it took an accidental discovery and the perseverance of two scientists to make it clear.

In 1969 a party of Japanese scientists working in Antarctica found nine meteorites lying on bare, blue ice near the Yamato Mountains in Queen Maud Land, Antarctica. Initially it was assumed that these were from a single meteorite shower, but when detailed mineralogic studies revealed that these were unrelated meteorites, light bulbs flicked on in

Figure 3.3 Recovery of meteorites from hot and cold deserts. Upper left: Meteorite man John Schutt on the trail in southeastern New Mexico. Upper right: Bill Cassidy, godfather of Antarctic meteorite recovery expeditions organized in the US. Lower right: Meteorite searching in Namibia – trying to figure just where the heck we are in pre-GPS days. Lower left: A typical hot desert meteorite lying as found, with my knife for scale.

the heads of a few meteoriticists. Bill Cassidy (University of Pittsburgh) and Keizo Yanai (National Institute of Polar Research in Tokyo) immediately realized that the special environmental conditions in Antarctica were preserving and concentrating meteorites to an astonishing scale. Some meteorites have been sitting on the ice awaiting discovery for two million years. A regular program of expeditions was initiated by the Japanese, under Keizo Yanai, and in the USA by Bill Cassidy. These efforts continue to the present day. From time to time other groups also organize special expeditions to the ice. After 25 years of collecting there are now upwards of 30,000 meteorites in our collections found in the Antarctic, approximately half of the entire world supply!

I have participated in two of the US expeditions (officially called the Antarctic Search for Meteorites (ANSMET). These were tremendous experiences that I would recommend to anybody who can tolerate a rather low comfort level. First, you travel to Christchurch, New Zealand, where you an outfitted with special polar clothing and equipment. Next you are transported to the main "town" in Antarctica, McMurdo Base in South Victoria Land, a seven hour flight over some of the Earth's most treacherous ocean. McMurdo has many of the aspects of a wild-west mining town. But this is plush compared to your accommodations in the field, where you can expect to spend up to seven weeks living in a two-man tent, hundreds of miles from the nearest TV-VCR, fresh food, heater, and toilet.

For those who appreciate such spartan conditions, however, the rewards are great. Clean air, beautiful scenery (as long as you like lots of rocks, ice, and eternally frozen wind reducing your hands and face to popsicles), and the excitement of new discoveries every day. Meteorite recovery parties traverse the ancient, blue ice fields in ski-doos, constantly on the lookout for celestial treasures. In field areas far from rock outcrops (and 99% of Antarctica is buried under ice up to 10,000 feet thick), every rock you see must be a meteorite. The ice movement serves to concentrate meteorites in occasional small areas, called accumulation zones.

I have seen a hundred meteorites in areas as small as an average house lot in Houston. In some years field parties return to civilization with over 1,000 new meteorites. The Japanese just returned with over 4,000, collected in a single field season.

One of the best things about these expeditions is the opportunity to experience this unique continent with smart, wonderful friends. For many years the expeditions were led by Bill Cassidy, an adventurous man whose career was shaped by many years of expeditions to truly remote places in search of meteorites, tektites, and meteorite craters. He has now retired and passed his Antarctic empire on to one of his final students, Ralph Harvey (of Case Western Reserve University), who has taken the program in new directions and into a new millenium. In all these 25 years not a single scientist has been killed, and only two have ever been seriously injured, and this is due solely to the vigilant efforts of the program's ice guide John Schutt. John is a true legend in our field – immediately likable, supremely capable, indestructible, and yet quite a daredevil. I believe that John must also have personally found more meteorites than anyone in history. We once went for a walk in a New Mexico desert and found two meteorites. Just walking around; only John could do that.

There is another kind of desert that has been providing us with many new meteorites, and these are hot deserts. Starting in the 1960s, an avid indian arrowhead collector in Portalis, New Mexico, Ivan "Skip" Wilson, began thinking about meteorites as his supply of artifacts began to dry up. He noticed that in basins where the wind had recently blown away fine sand (deflation basins), the accumulated rocks sometimes included meteorites. Over the succeeding three decades, Skip has found hundreds of meteorites in Roosevelt County, New Mexico. This experience jolted others into searching their own favorite deserts, and today there are many individuals recovering meteorites daily from the Sahara, Namib, Nularbor, Sonora, and other hot deserts. I have personally participated in searches in Namibia, Chile, and the southwestern USA. It is clear that many more meteorites await discovery in deserts all over Earth.

In fact, new meteorite finds are flooding in so rapidly that it is almost impossible for meteoriticists to keep up with classifications and names. The primary benefit from this sudden wealth is the many unusual and unique meteorites that have appeared. Bill Cassidy always refers to the ANSMET Program as a very cheap sample return space mission. Everyone is aware of the Lunar and Martian meteorites we now have for study. All of the former, and the majority of the latter meteorites have been found during expeditions to the hot and cold deserts, including the first ones to be definitively identified. Where would our understanding of Mars be today without these Martian meteorites? It is undeniable that the claims of the discovery of fossils in a Martian rock, found in Antarctica in 1984, have been a major driver for the current robust Mars exploration program being carried on by every space-faring nation. The fact that practically no scientists believe the fossil claims is beside the point. These Martian meteorites are now at the heart of our knowledge of Martian geology, geochemistry, formation and subsequent geological history, atmospheric studies, and on and on. Of course, these Martian meteorites will never answer all of the critical questions we have about Mars, only carefully executed sample return missions will do that.

Watch the skies!

The US has placed a constellation of DOD satellites in orbit whose solitary duty is to watch for launches of nasty ICBMs against our country. In the 30 years over which they have been on guard they have fortunately detected no hostile missile launches. However, they have observed hundreds of large meteoroids entering the Earth's atmosphere. It is critically important to determine the identity and, if possible, source of this entering material, so that we can monitor the cosmic debris raining on to our planet. Unfortunately, although we have the satellite detections, we have never successfully recovered fireball dust in the atmosphere or meteorites on the ground after a detection. Still we knew if we were patient our luck could change.

On January 18, 2000, sensors aboard two of these DOD sentinels detected the entry of a huge meteoroid into Earth's atmosphere over western Canada. Based on the luminosity of the fireball, Peter Brown (University of Western Ontario) calculated that this was the largest object to enter our atmosphere in a year, and the largest over land in a decade. The energy released by the fireball in the two seconds it took to traverse the atmosphere was equivalent to two to three kilotons of TNT. Residents of Whitehorse in Yukon Territory were treated to an intense dawn fireball, thunderous booming, and sizzling sounds, a sulfurous odor, and a luminous cloud that persisted for up to an hour. Many individuals reported a metallic taste in their mouths, which is still unexplained. The huge dust cloud that had been produced told us that this was a very friable meteorite – the sort of material that almost never reaches the ground or survives there for long.

For this exceptional event we pulled out all the stops, making every possible effort to recover material for analysis. Within 48 hours we sent a NASA stratospheric ER-2 aircraft (based in southern California) to western Canada in an effort to collect any residual dust that might remain in the air from the fireball. Because 48 hours passed before we could do this, however, the chances of success were slim.

Realizing this sad fact, we requested that local residents collect snow samples from frozen lakes, in the dim hope that fallout from the event would be present. Because of the time of year (winter) and sparsely populated nature of the target area we had no reasonable expectation that actual meteorites might be recovered. We were wrong. One week after the fireball, local resident Jim Brook was driving his pickup across frozen Tagish Lake to his home in the bush. He was forced off of his normal track by some pressure ridges in the ice. As he picked his way around this barrier he noticed some dark objects scattered across the frozen surface. He quickly realized that he had found a fortune lying practically in his front yard. Over the next day he collected one kilogram of samples, and he had the foresight to keep the samples frozen. A snowfall brought a sudden close to his treasure hunting. Alan Hildebrand (of the University of Calgary) led an expedition to attempt further meteorite recovery in the snow. Even when an enthusiastic drug-sniffing Royal Canadian Mounted Police dog was brought on to the case, and given the scent of a meteorite, no new specimens were found. Collection activities were not resumed until the May spring thaw.

At the urging of Peter Brown and Canadian Survey Geologist Charley Roots, Jim sent half of his cosmic harvest to me, by frozen express air shipment, for initial characterization. US Customs agents almost would not release the samples to me, so unusual was the package. As I argued with them I could not help thinking that this was probably not a meteorite, and that I was wasting an afternoon. You can imagine my feelings when I finally opened the crate back in my lab and realized that not only was this indeed a meteorite, but a rare carbonaceous chondrite. These particular meteorites are a special love of mine, and I had always believed that I would only have this experience in wild dreams. The last major fall of a carbonaceous chondrite occurred when I was attending Denonville Junior High School, in 1969. Even now I can scarcely believe this really happened. As we began analyzing the Tagish Lake meteorite, we only gradually realized just how special it really was.

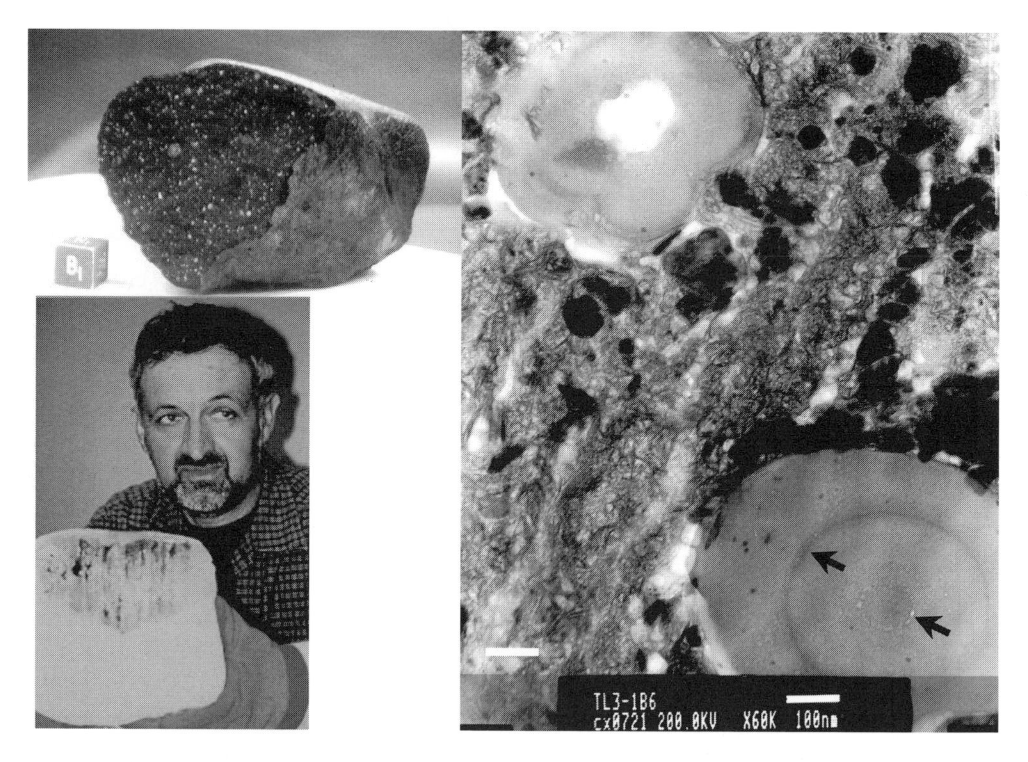

Figure 3.4 Tagish Lake meteorite fall and analysis. Upper left: A sample of a frozen, pristine Tagish Lake stone in the lab. Lower left: Jim Brook holding a block of ice containing small meteorites recovered in the spring. Right: A transmission electron microscope image of hollow organic globules present within the Tagish Lake meteorite. The scale bar measures 100 nanometers; arrows point to layers within a single globule. Photo courtesy of Keiko Nakamura.

With the spring thaw in May, Alan Hildebrand and Peter Brown led a triage meteorite recovery effort in the rapidly melting ice-capped Tagish Lake. By the time the now disintegrating lake ice had become too dangerous to cross, their brave, wet team had collected hundreds of additional samples. Many of the samples were found frozen into lake ice, like tiny wooly mammoths. They found that most of the samples had turned to mud upon becoming wet, and were unrecoverable. If Jim had not recovered his samples so quickly, and before the first snow, they would have never survived intact.

The fall and recovery of the Tagish Lake meteorite has been an incredible bonanza for our field, and we will be reading its record of early Solar System secrets for many years to come. The recovery of the Tagish

Lake meteorite is also unique in that Jim Brook took the precaution of maintaining the first-recovered samples in a frozen state, as they remain to this day. Meteorites are cold-soaked in the deep freeze of space for millions of years, and though they heat up to incandescence for a few seconds in the atmosphere, only the outer skin actually heats up in this brief period. Once on the ground the hot skin quickly cools; inside the meteorite remains frozen and only reaches room temperature after some time. Still, people who recover meteorites a few seconds after their fall have never thought to put them into a freezer – but Jim Brook did. Because the Tagish Lake meteorite landed in a frozen landscape, and was placed into a freezer immediately upon recovery, it is the only extraterrestrial material that has ever been recovered frozen. Why is this is so important will soon become apparent.

Imagine being fortunate enough to recover pieces of such a huge object; this was truly wonderful. I now believe that this type of meteorite falls to Earth very commonly, but because it is so friable is simply never recovered on the ground for study. But, more surprises were to follow as we proceeded to the analysis of this meteorite. As we suspected from the reports of the foul odor attending the fireball, the Tagish Lake meteorite was a very special visitor, a carbonaceous chondrite. These meteorites are very rich in carbon and the organic precursors to life. Organic molecules as complex as amino acids are present. Water is present within clay minerals and other products of aqueous processes on asteroids. Many of us believe that meteorites like Tagish Lake provided the Earth with much of its water and organics – the most common molecules in your body. This is why it was so fortunate that this meteorite was maintained in a frozen state – very volatile organic molecules that always leak away as meteorites heat up are, in this one case, still trapped inside.

Currently Sandra Pizzarello (Arizona State University) and Iain Gilmour (Open University) are sorting through the organics in Tagish Lake, and it turns out that these are very strange – not at all what we expected. Compared to similar carbonaceous chondrites, the quantity and diversity of amino acids in Tagish Lake is very limited. Water-soluble organics are almost entirely lacking. Luann Becker (then at the University of Washington) further reports the presence of an unusual suite of fullerene molecules (caged structure compounds consisting solely of carbon). Until the results of further work are available it is not possible to understand the story that these organic molecules are telling. However, two scenarios are now equally possible. In the first scenario Tagish Lake organics are much more primitive, and therefore less diverse, than those in other meteorites, reflecting a very ancient and pristine character – a sample of the initial organic population in the Solar System. However, it is also possible that Tagish Lake developed the typical suite of complex organics, like those found in other carbonaceous chondrite meteorites, but then some chemical event erased much of the record. Water could have done this, in a process similar to what happens during the rinse cycle in a washing machine.

In a stunning development, Keiko Nakamura (Kobe University), a young mineralogist working with our group ever since she was a college summer intern, has recently found in one Tagish Lake sample microscopic hollow hydrocarbon globules. While we see no evidence of true biologic materials, these globules would have been made-to-order structures for early organisms. We have never seen these in any meteorite, but such objects have been produced in the lab by Dave Deamer and associates at NASA's Ames Research Center, in the course of experiments simulating interstellar reactions. Who knows what we will find as we continue our exploration of the organics in this unique meteorite?

Another very exciting but enigmatic feature of the Tagish Lake meteorite is the presence of a huge quantity of calcium, iron, and magnesium carbonate minerals. On Earth such minerals commonly form from the evaporation of huge quantities of groundwater or seawater. Again, here is an indication that the Tagish Lake home asteroid was an exceptionally wet world. These carbonates also hold the promise that they may contain aqueous fluid inclusions, a common feature of carbonates on Earth, as well as providing the means of age-dating the aqueous alteration events out in the asteroid belt.

Carbonaceous chondrites also contain preserved presolar grains, which are direct samples of the very stars that provided all of the atoms in the Solar System, and it turns out that Tagish Lake contains these in record quantities. We now believe that this meteorite comes from somewhere out in space near Jupiter, as a piece of a D class asteroid. Imagine finding something from so far out in space by looking down to the frozen ground at your feet!

Future directions

Being originally a field geologist and mineralogist, I really want to know what it is like to walk across the asteroids and comets, kick up some rocks and dust, stoop down and take a really close look. Of course I was born rather too early to have a shot at this, but I can do the next best thing. I have joined the science teams for the first two sample return missions from a comet and an asteroid – the *Stardust* mission launched in 1998. It will pass by the short-period comet Wild II on January 1, 2004, and return to the Earth in early 2006. During the comet flyby the spacecraft will open to expose a tray of silica aerogel blocks, which will capture and preserve thousands of comet coma dust grains as they impact. Silica aerogel is a magical transparent substance that is only slightly more dense than air, and has proven capabilities to capture grains fired into it at high velocities. This will be no small achievement as the comet will pass by the Stardust spacecraft at a velocity of over 6 km/sec – about ten times the velocity of a rifle bullet. The Wild II samples will be our first real look at material that we believe dates from the very dawn of our Solar System.

I am also fortunate to be on the science team for the world's first sample return mission from an asteroid. The MUSES-C mission, scheduled for launch in 2002, is primarily a Japanese mission, but with some support from NASA scientists and engineers. The MUSES-C spacecraft will rendezvous with asteroid 1998SF36, descend and gently bounce off the surface. During this touch-and-go maneuver a small gun will fire into the surface, liberating chips and dust, which will be stored aboard the spacecraft for Earth return in 2007.

These are the first sample return missions to be attempted in 30 years, since the Apollo and Luna missions to the Moon. Those previous missions returned large amounts of rocks and soil – samples any geologist could see, hold, and identify with. By contrast, the new generation of sample return missions are going to be returning very tiny, sometimes even microscopic samples. Fortunately, one of the dominant trends in meteoritics has been learning to do more (a lot more) with vastly smaller samples. Many older NASA employees speak of the Apollo days as the high point of their lives, and seem to feel that nothing as exciting will happen again. I disagree. It is clear that we are only now entering a new golden age of sample return missions, for which our studies of meteorites have prepared us well. The most exciting discoveries are yet to be made.

Further reading

McSween, H. Y. Jr. (1993). *Stardust to Planets*, New York: St. Martin's Griffin, 241p.

Norton, O. R. (1994). *Rocks From Space,* Missoula, MT: Mountain Press Publishing Company, 449p.

Dodd, R. T. (1986). *Thunderstones and Shooting Stars*, Cambridge, MA: Harvard University Press, 196p.

4

Mercury: the inner frontier

Robert G. Strom, University of Arizona

Bob Strom, a native of California, graduated from Stanford in Geology in 1957 and has made his life's work the study of the terrestrial planets. But among the inner planets, Bob's favorite is Mercury. Indeed, Bob is perhaps the world's biggest planet Mercury lover (he's a Mars lover too, but Mercury is his deepest planetary passion). A researcher in planetary sciences since the early 1960s, at the very dawn of planetary exploration, Bob has mentored students, participated in missions of exploration, and written numerous research papers. His book, *Mercury: The Elusive Planet,* is a wonderful read that I cannot recommend highly enough. He has one son and two grandchildren. He and his wife Ayako live in Tucson. Bob enjoys traveling and lecturing for cruise lines.

In a very large family there is usually at least one child that is neglected for one reason or the other. The child may not be as handsome or beautiful, or may not seem as interesting or to have the potential of the other children. In the family of planets that child is the little runt with the pockmarked and wrinkled face named Mercury. There has been only one mission to Mercury and earth-based telescopic observations are extremely difficult. As a consequence we know less about Mercury than most other planets in the Solar System. Our knowledge of the geology, surface composition, interior, magnetic field, exosphere, polar deposits, thermal history, and origin is poor to extremely poor.

I first became involved with Mercury when I became a member of the Mariner 10 mission to Venus and Mercury in 1970. At that time there was almost nothing known about this innermost planet except its orbit, mass, and that it was smaller than other planets. Because it is so close to the Sun, its speed is faster, and, therefore, its orbital period is shorter than any other planet. It makes one orbit around the Sun in only about 88 Earth days. It has a peculiar orbit with the largest eccentricity and inclination of any planet except Pluto. The inclination of the orbit is about 7 degrees and the eccentricity is about 0.205. This means that Mercury's orbit is so elliptical that its distance from the

Sun varies by about 34%. We also knew that Mercury's surface is very hot. Its sunlit side is about 427 degrees C, and its night side is about −183 degrees C. Therefore, Mercury has the greatest range in temperature (610 degrees C) of any body in the Solar System. Observing Mercury is extremely difficult because it is so close to the Sun. It is never more than 28 degrees from the Sun as viewed from Earth, and, therefore, observations must be made through a long path length of Earth's atmosphere. This causes terrible observing conditions.

In 1970, we knew nothing about its surface except that it reflects about the same amount of light as the Moon. Its mass was fairly well known from its orbital characteristics, but its size was not too well known (4,878 kilometer diameter). However, its size was known well enough to understand that it had a very large density for such a small planet. Its density of 5.44 gm/cm^3 is larger than any other planet except Earth (5.52 gm/cm^3). Because of Earth's large internal pressures, however, its uncompressed density is only 4.4 gm/cm^3 compared to Mercury's uncompressed density of 5.3 gm/cm^3. This strongly indicates that Mercury has an enormous iron core about 75% of the planet's diameter (Figure 4.1, see color plate section). This core is proportionally larger than in any other planet. Why it has such a large iron core is still not well understood.

Mariner 10 observations

Although we learned a lot about Mercury from Mariner 10's three flybys, there were more questions raised than answered. The spacecraft only imaged about 45% of the planet and that was not seen in great detail; about the same detail as telescopic observations of the Moon from Earth. Furthermore, much of this coverage was at high sun angles where terrain analysis is difficult to impossible. Consequently, only about 25% of the surface was seen under sun angles favorable for terrain analyses and other geological studies.

Although it was known that Mercury was not in synchronous rotation before Mariner 10, the rotation rate was measured with great accuracy, at 58.646 days, from spacecraft observations. Since Mercury's orbital period is 87.97 days, it was realized that Mercury has a unique 3:2 spin-orbit coupling where it makes exactly three rotations on its axis for every two orbits around the Sun. This unique resonance results in an exceedingly long solar day (sunrise to sunrise) of 176 days or two Mercurian years. Apparently the resonance was acquired over time as the natural consequence of the dissipative processes of tidal friction and the relative motion between a solid mantle and a liquid core. Furthermore, it was discovered that the obliquity of the planet is close to 0 degrees. Therefore, it does not experience seasons as do the Earth and Mars. Consequently, the polar regions never receive the direct rays of sunlight and are always frigid compared to torrid sunlit equatorial regions.

Another effect of the 3:2 resonance is that the same hemisphere always faces the Sun at alternate perihelion passages. This happens because the hemisphere facing the Sun at one perihelion will rotate 1.5 times by the next perihelion, placing it directly facing the Sun again. The subsolar points at 0 degrees and 180 degrees longitudes occur at perihelion so they are called "hot poles," while the subsolar points at 90 degrees and 270 degrees longitude occur at aphelion and are called "warm poles."

Mariner 10 discovered that Mercury has a global intrinsic dipole magnetic field similar to Earth. Neither the Moon nor any other terrestrial planet currently has an active dipole magnetic field. The magnitude of the dipole moment is about 330 nT, or over 1,000 times smaller than Earth's. Although weak, it is strong enough to hold off the solar wind, creating a bow shock, a magnetosheath, and a magneosphere. Its magnetosphere is essentially a miniature version of Earth's magnetosphere; about 7.5 times smaller. However, because of the weaker field, Mercury occupies a much larger fraction of the volume of its magnetosphere than do other planets, and the solar wind actually reaches the surface at times of highest solar activity. As a consequence of its small size, the equivalent regions of intense radiation belts in the Earth's and the outer planet's magnetic fields are below the surface of Mercury. Furthermore, magnetic events happen more rapidly and repeat more often than in Earth's magnetosphere. Although other models may be possible, the maintenance of terrestrial planet magnetic fields is thought to require an electrically conducting fluid outer core surrounding a solid inner core. Therefore,

Figure 4.2

The incoming side of Mercury as viewed by Mariner 10.

Mercury's magnetic field is taken as evidence that the outer core is still molten. Since Mercury is so small (only about one-third the size of Earth), it should have lost enough heat in the last 4.5 billion years to completely solidify the core unless there was some low melting point substance in the core. The most likely substance is sulfur, but sulfur should not be there if Mercury formed at its present distance from the Sun. So what's going on? We will come back to this later.

Although Mercury does not have an atmosphere it does have an exosphere where atoms rarely collide. Mariner 10 discovered hydrogen, helium, and oxygen and set upper limits for argon, and later Earth-based measurements discovered other more abundant constituents, sodium and argon.

Mariner 10 showed that the surface of Mercury has been heavily cratered by impacts (see Figures 4.2 and 4.3). However, there are wide-

Figure 4.3

The outgoing side of Mercury as viewed by Mariner 10.

spread regions of plains, called intercrater plains, between clusters of craters. In fact, this is the most common terrain in the area imaged by Mariner 10. Filling and surrounding large impact basins, such as the 1,300 kilometer diameter Caloris Basin, are smooth plains that are thought to have a volcanic origin (see Figure 4.4). The floor of this impact basin has a tectonic structure unlike any other basin in the Solar System (see Figure 4.4). It consists of closely spaced ridges and troughs arranged in both concentric and radial patterns. The ridges are probably caused by compression, and the troughs are probably tensional graben, which post-date the ridges. This pattern of tectonic features is consistent with subsidence and subsequent uplift of the basin floor. Directly opposite the Caloris basin on the other side of Mercury is a peculiar broken-up region called the hilly and lineated terrain (see Figure 4.5). This terrain is thought to be due to titanic seismic waves created by the

Figure 4.4

A Mariner 10 mosaic of the 1300 kilometers diameter Caloris impact basin. Smooth plains fill and surround the basin. Notice the peculiar interior structure of the basin floor.

massive Caloris impact and focused at the opposite point of impact. Computer simulations suggest that the vertical motion of the surface due to the seismic waves could have been greater than two km.

Very little is known about the surface composition of Mercury. Mariner 10 did not carry instruments that would provide good information on its composition, although it did image the surface through a variety of filters that could be used in infer something about its composition. If the plains units are volcanic in origin then they must be comprised of very fluid lava with viscosities similar to those of fluid flood basalts on the Moon, Mars, Venus, and Earth. The photometric characteristics are very similar to those of the Moon. However, at comparable phase angles and wavelengths in the visible part of the spectrum, Mercury appears to have systematically higher albedos than the

Figure 4.5

This broken-up terrain is on the other side of Mercury directly opposite the Caloris impact basin (the antipodal point). It is thought to be the result of seismic waves from the impact, and that were focused at the antipodal point. The image is 540 kilometers across.

Moon. Normal albedos range from 0.09 to 0.36 at a 5-degree phase angle. Although the higher albedos are usually associated with rayed craters, the highest albedo (0.36) on the Mariner 10 images is the floor deposits of Tyagaraja crater at 3 degrees north latitude and 149 degrees longitude. The lunar highlands/mare albedo ratio on the Moon is almost a factor of 2, whereas the smooth plains/highland ratio on Mercury is only about 1.4. At ultraviolet wavelengths (58 to 166 nm), Mercury's albedo is about 65% lower than the Moon's at comparable wavelengths. These differences in albedo suggest that the composition of the surface of Mercury is different than that of the Moon.

Mark Robinson at Northwestern University has recalibrated and color-ratioed the Mariner 10 images to derive new compositional information on the FeO (iron oxide) abundance, the opaque mineral content, and the maturity of the regolith. He finds that the probably volcanic smooth plains have a FeO content of <6% weight, which is similar to the rest of Mercury imaged by Mariner 10. This suggests that Mercury's surface has a more homogeneous distribution of elements affecting color (e.g., more alkali plagioclase) than does the Moon. At least the smooth plains may be low-iron alkali basalts. The iron content of lavas is thought to be representative of their mantle source regions. Therefore, Mercury's mantle may have the same FeO content ($< 6\%$ weight) as the crust, indicating that Mercury is highly reduced with most of the iron in the core. This contrasts sharply with the estimated FeO content of the bulk Moon ($\sim 11.4\%$), of Venus and the Earth ($\sim 8\%$), and of Mars ($\sim 18\%$).

Associated with some fractures are dark blue, low-albedo, and high-opaque mineral regions that could be more mafic volcanic pyroclastic deposits. Bright rays have a very low opaque mineral content, which could indicate the impacts penetrated into an anorthositic crust. Color ratios between lunar and Mercurian rays suggest that Mercury's surface is low in $Ti4^{+}$, $Fe2^{+}$, and metallic iron compared to the Moon's surface. Earth-based microwave and mid-infrared observations also indicate that Mercury's surface has less FeO plus TiO_2, and at least as much feldspar as the lunar highlands. Although this has been interpreted to mean that Mercury's crust is devoid of basalt, it could just as well mean that the basalts have a low iron content or are fluid alkali basalts. On Earth there are low-viscosity alkali basalts that could produce the type of volcanic morphology seen on Mercury's plains.

During the Mariner 10 encounters with Mercury, I noticed that there were numerous lobate scarps that occurred on almost every image of Mercury (see Figure 4.6). I became intrigued with their origin and what they could tell us about the tectonic and thermal history of the planet. They were similar to wrinkle ridges on the Moon that are largely confined to the lunar mare regions, and which had been interpreted as thrust faults. However, Mercury's ridges are generally larger, and they are not confined to any particular terrain. Furthermore, one scarp definitely showed that they must be thrust faults due to compressive stresses. This particular scarp displaced a crater rim and shortened its radius in such a way that it could only be caused by compression. I set out to map them, measure their length, estimate their height from shadow measurements, and establish their age relative to other features. To my surprise they were distributed more or less randomly over the surface of Mercury favorably viewed by Mariner 10.

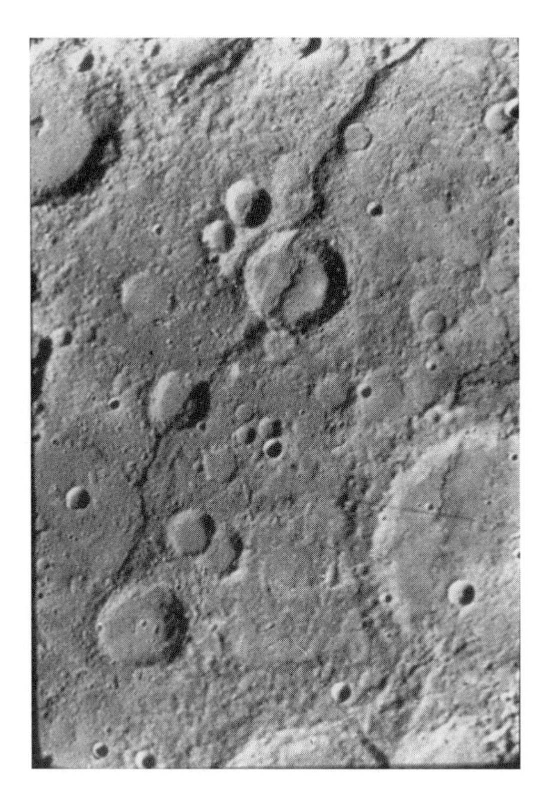

Figure 4.6

One of the lobate scarps that are probably thrust faults caused by compression due to planet cooling. This one is named Discovery Rupes. It transects two craters 55 and 35 kilometers in diameter.

It was not possible to measure the height of every scarp, but ones near the terminator that have cast shadows were about 200–500 meters high. One, named discovery scarp, was estimated to be over one kilometer high. Some were over 500 kilometers long and all appeared to be younger than the intercrater plains. Very few impact craters disrupt scarps and they transect both subdued and fresh craters. These observations suggested that they were formed relatively late in Mercurian history. Another important observation was that the azmuthal orientation of the scarps seemed to be random. Estimates of the fault plane inclination based on terrestrial thrust faults, their lengths, and number, together with an estimate of the average scarp height, suggested that if the scarps were distributed in a similar manner on the unseen half to Mercury then there was a decrease in the planet's radius of about one to two kilometers.

The most logical explanation for these observations is that Mercury was subjected to a period of global compression relatively late in its history. This compression could be the result of cooling of the planet from an originally hotter state. Some thermal models, in fact, predict a contraction of that magnitude. However, since we have not analyzed more than about 25% of the surface, these conclusions have to remain very tentative until more information is available from future missions. Since this research, I have concentrated on trying to understand the history of Mercury and what that may tell us about the evolution of the terrestrial planets. Our general understanding of Mercury's geologic history from the Mariner 10 data is that its earliest recorded surface history began after core formation and a possible mantle-stripping event (see section on Mercury's origin).

The earliest events are the formation of intercrater plains > 4 billion years ago during the period of late heavy bombardment. These plains may have been erupted through fractures caused by large impacts in a thin lithosphere. Near the end of the bombardment, the Caloris Basin was formed by a large impact that caused the hilly and lineated terrain from seismic waves focused at the antipodal region. About 3.8 billion years ago lava erupted within and surrounding the Caloris and other large basins to form the smooth plains. The system of thrust faults formed after the intercrater plains due to cooling of the planet. This resulted in about a one to two kilometers decrease in Mercury's radius. The core continued to cool and the lithosphere thickened causing more compression and closing off of the magma sources. Volcanism apparently ceased near the end of the period of late heavy bombardment. Today the planet may still be contracting as it continues to cool. This general history is probably over simplified, because we have not seen 55% of the surface at all, and 25% was viewed at unfavorable sun angles so that geologic analyses could not be done.

Thanks to Mariner 10 we learned that Mercury had a heavily cratered surface that represents a period of intense bombardment early in Solar System history, and that there are plains that may be the result of very early volcanism. We also knew that the tectonic framework was different from any other planet or satellite, and that Mercury had a magnetic field and probably a liquid outer core of unknown thickness. This raised questions about Mercury's origin that affect our understanding of the entire Solar System. In the 27 years since Mariner 10 there have been some observations from Earth-based facilities that have made Mercury even more interesting. The Mariner 10 mission was never meant to be the definitive mission to Mercury. In fact, it was conceived as only a first reconnaissance mission to see what was there, and to define experiments for a more sophisticated orbiter to follow. However, 28 years after its first encounter with Mercury, Mariner 10 remains the only spacecraft to have visited the planet.

I think there are two main reasons why it has been such a long time between Mercury missions. First, placing a spacecraft in orbit around Mercury is not easy. Mercury's close position to the Sun makes for a very severe environment. The high temperatures and a high flux of energetic particles from the Sun can damage the spacecraft and its electronics. Therefore, the spacecraft must be well shielded against such radiation. Also it takes a long time to get there if you want to put a spacecraft in orbit. A reasonable launch vehicle, a realistic amount of onboard fuel, and a modest retro-rocket require that the spacecraft encounter Mercury at a relatively low velocity in order to put it into orbit. If the spacecraft was launched directly to Mercury from Earth its encounter velocity would be so high that it would require a retro-rocket the size of a launch vehicle to slow it down enough to orbit the planet. Consequently, the spacecraft must encounter both Venus and Mercury multiple times to gravitationally slow the spacecraft enough to achieve orbit. This takes about five years. A lot can happen to a spacecraft in five years, and none of it is good. Because of these difficulties an orbital mission to Mercury is necessarily expensive and time consuming.

The second reason there has not been a mission to Mercury in the last 28 years is that the images of Mercury from Mariner 10 showed a heavily cratered surface similar to that of the Moon. These images did not reveal enormous shield volcanoes, deep canyons, and huge flood channels as occur on Mars, nor did they show the huge fracture belts, large volcanoes, and peculiar surface features that occur on Venus. Most people thought it was just a dead Moon-like planet, and, therefore, not as exciting as the other bodies in the solar system and a low priority for further exploration. However, Mercury is very different from the Moon, as we shall see below.

The Moon and Mercury

In thinking about what we say of Mercury from Mariner 10, I am reminded of the flights of Mariners 6 and 7. They flew by Mars in 1969 and imaged a small portion of the surface that happened to be heavily cratered. Because it looked like the Moon people thought that Mars was dead and very similar to the Moon. Fortunately, Mariner 9 was launched soon afterward and showed from orbit that Mars had many diverse features totally different from the Moon. Could that also be true of Mercury? No planet or satellite shows a surface that has a uniform distribution of geology. Why should Mercury be different?

Is Mercury really as similar to the Moon as some people believe? I have listed in Table 4.1 the similarities and differences between the Moon and Mercury. As you can see, the differences far outweigh the similarities. The only similarities are the heavily cratered surfaces, the smooth plains and the superficial surface layer called a regolith (a kind of soil resulting from the grinding-up of material by impacts of particles of all sizes). The only reason this layer is on the Moon and Mercury is because they both lack substantial atmospheres to shield the surface from smaller impacts. Heavily cratered surfaces occur on Mars and smooth plains occur on both Mars and Venus so there is no great difference there. But probably the major difference is that Mercury has an enormous iron core, and the Moon has hardly any. This is almost certainly due to their very different origins. Mercury also has a dipole magnetic field. In fact, other than Earth, Mercury is the only terrestrial planet with a currently active magnetic field.

Table 4.1. *Mercury/Moon comparison*

Similarities to the Moon
1 Heavily cratered surface
2 Smooth plains associated with impact basins
3 Regolith (impact produced surface layer)

Differences from the Moon
1 Large iron core ~75% of the diameter
2 Magnetic field
3 Large areas of inter-crater plains (the major terrain type)
4 Comparable geologic units are brighter
5 Widespread distribution of thrust faults (unique to Mercury)
6 Unique impact basin floor structure (Caloris)
7 Unique radar feature
8 Very strong radar signature from polar deposits
9 Origin

Mercury's surface is dominated by intercrater plains that are found (not surprisingly) between clusters of craters. Although the Moon has patches of such plains they are very small. The tectonic framework is not only different from the Moon, but it is different from all other bodies in the Solar System. Also the crater rays and other comparable lunar features appear to reflect more light than on the Moon. A recent re-evaluation of color information derived from Mariner 10 images suggests that Mercury's composition is different than that of the Moon. There are also radar bright deposits in the polar regions of Mercury that are similar and as pure as the ice deposits on the Ganymede, an icy satellite of Jupiter. One radar anomaly is different from anything else in the Solar System. Also the floor structure in the Caloris basin is different from any other impact basin floor structure in the Solar System, and the abundances of elements found in the exosphere are very different from those in the Moon's thinner exosphere. Finally, the origin of the two objects is completely different. The Moon was the result of a large impact with the Earth very early in its history, but Mercury must have had a radically different origin. I think it is entirely obvious that Mercury is not just another Moon-like body. It has characteristics that are not only radically different from the Moon, but different from other Solar System bodies. Post Mariner 10 observations of

Mercury make the planet even more interesting than previously believed.

Recent observations from Earth

In 1985–1986, Earth-based telescopic observations detected for the first time sodium (Na) and potassium (K) in the exosphere of Mercury. Sodium is the most abundant element. Both sodium and potassium have highly variable abundances (10,000–10,000 Na atoms/cm^3 and 100–10,000 K atoms/cm^3) near the surface on time scales of hours to years. Their abundances also vary between day and night by a factor of about 5, the dayside being greater. Both of these elements probably originate from the surface, but we are not sure how they are derived or what this means about the composition of surface materials. Also some areas are apparently associated with enhancements of sodium and potassium, such as the Caloris Basin. Photo-ionization of the gas will result in the exospheric ions being accelerated by the electric field of Mercury's magnetosphere.

The net result is that ions will be ejected in to space and lost. The loss rate for Na atoms is about 1.3×10^{22} atoms per second. Therefore, atoms must be continuously supplied by the surface. The total fraction of ions lost to space is at least 30%. Therefore, the atmosphere is transient and exists in a steady state between its sources and sinks. Sodium and potassium are probably derived from the surface of Mercury, but the mechanism that supplies them is not well understood. They could be released from sodium- and potassium-bearing minerals by interactions with the solar wind, or they could be supplied by impact vaporization by micro-meteoroid bombardment.

Other elements are probably present but remain undetected. In 1991, high-resolution, full-disk radar images of Mercury showed polar features with high reflectivities and polarization ratios. These radar characteristics are similar to water ice on the Galilean satellites of Jupiter and the residual polar water-ice cap on Mars. Furthermore, the deposits are confined to the permanently shadowed regions of craters where temperatures are low enough to preserve water ice for geologically long periods of time. The high radar reflectivities and circular polarization inversion are consistent with volume scattering in relatively thick, clean ice that has been cold trapped in the permanently shadowed regions of craters. However, sulfur has also been suggested. The south polar feature is centered at about 88 degrees south and 150 degrees west in a crater (Chao Meng-Fu) that is 150 kilometers in diameter. Recent very high-resolution (1.5–3 kilometers) radar images of the north polar regions have detected these deposits in craters as small as ten kilometers diameter and at latitudes as low as 72–75 degrees. The Moon does not show similar radar anomalies in the permanently shadowed regions of craters near the poles. However, relatively high abundances of hydrogen detected in lunar polar regions by the Lunar Prospector mission have been interpreted as indicating ice. If this is the case the water ice is more likely to be in the form of dilute frost mixed in with the regolith than the relatively thick deposits of clean ice on Mercury. The water could be external in origin and derived by comet and/or water-rich asteroid impacts. Bright radar features in the southern and northern hemisphere are probably fresh impact craters. Possibly these features are the result of recent comet or water-rich asteroid impacts that supplied the polar deposits.

In early 1973, several months before the Mariner 10 launch, I was attending a Mariner 10 meeting at JPL in Pasadena, California. At the meeting Richard Goldstein presented some results of observing Mercury with the large radar dish at the Goldstone facility in the Mohave desert. He had discovered three radar-bright spots on Mercury's surface (not the polar deposits discovered later). The observations did not show much detail, but they could be important to understanding the surface evolution of the planet. When he told us where they were located, we were disappointed to learn they were on the side we would not see from Mariner 10. About a year later when Mariner 10 flew by Mercury and showed a heavily cratered surface I thought these bright radar features were nothing more that relatively fresh impact craters. Fresh impact craters on the Moon give radar-bright signatures.

I did not give these features a further thought until 23 years later at a special Mercury session of the 1996 COSPAR meeting in Birmingham,

England. At this special session, which was held to report on new Mercury data in preparation for a possible European Mercury mission, John Harmon presented results of radar observations with the very large radar dish at Arecibo in Puerto Rico. These radar images were at a resolution of about 15 kilometers, and showed that Feature A was an impact crater. However, Feature B looked more like a shield volcano and Feature C was not interpretable. In 1999 John re-imaged the surface at much high resolution (~1.5 kilometers) using the newly resurfaced Arecibo radar dish. Both Features A and B have all the radar characteristics of large fresh impact craters; no surprise there! However, Feature C is very different. All we can say about this feature is that it has a radar signature characterized by a cluster of bright spots collectively over 1,000 kilometers in diameter. It is certainly not an impact crater. In fact, it is unlike any other radar feature in the Solar System. What it could be is anyone's guess. We will just have to wait for Messenger to tell us what this intriguing feature is.

Mercury's origin

One of the outstanding problems in planetary science is the manner in which Mercury acquired such a large iron core. In other words, how did the planet originate? Early cosmochemical models of Solar System formation predicted that if Mercury formed at its present distance from the Sun, it should not have such a large iron core, and there should be no sulfur and practically no water in the planet. Without sulfur it is almost impossible to keep any part of the core in a molten state over geologic time as suggested by the dipole magnetic field. However, later mechanical models of inner planet formation suggested that a significant portion of Mercury could have formed from material derived from farther out in the inner planet zone, where sulfur and water are stable. According to the model some of this material is forced inward by gravitational forces to help form Mercury. Large bodies can also form which may be forced into orbits that cross the newly forming terrestrial planets. These can collide with the newly formed planets and result in major disruptions, including complete fragmentation of the objects. In fact, two such collisions probably created the Moon, and reversed and slowed the rotation of Venus. Although this model could account for the sulfur to keep Mercury's core partly molten, it still cannot account for its large iron core. It only results in an uncompressed density of 4.2 gm/cm^3, rather than the observed 5.3 gm/cm^3.

There are three main theories to account for Mercury's large core. One proposes that the large core is the result of a concentration of iron due to dynamical sorting of iron from silicates and the preferential loss of silicates in the denser solar nebula at Mercury's distance from the Sun. One problem with this idea is that sulfur should be absent, and, therefore, the core should be completely solidified; contrary to indications, from the presence of a dipole magnetic field. A second proposal is that a high-intensity phase of the Sun early in its history, called a T Tauri phase, vaporized and drove off much of the silicates in the crust and mantle of a much larger Mercury, leaving behind a large iron core surrounded by only a thin silicate mantle. The third idea is that a large impact of a planet-sized object literally blasted off much of the silicate material. In this computer model, the entire planet is disrupted, but in the reaccretion process a large part of the silicate portion is lost when it is drawn into the Sun, but there is little loss of the iron.

In the first hypothesis Mercury was formed with a large iron core, and in the other two models Mercury was once larger with a normal-sized core but had its outer silicate layers stripped away. Fortunately, each hypothesis predicts a different composition for the silicate mantle. Today we cannot measure the composition with sufficient accuracy from Earth-based observations to decide between these models.

Mercury's future exploration

The era of not having good data for Mercury, to help decide among the many competing hypotheses, is almost over. Not only will there be a new mission to Mercury, there will be two. NASA has approved a Discovery mission called MESSENGER to orbit Mercury. MESSENGER is an acronym for MErcury, Surface, Space ENvironment, GEochemistry, Ranging. It is an appropriate name because Mercury is the Roman messenger of the gods and the acronym indicates the type of studies that will be carried out.

MESSENGER will be launched in March 2004, make two flybys of Venus (2004 and 2006) and two of Mercury (2007 and 2008), before it is placed into orbit in April 2009. It will carry a complement of instruments to image the entire surface in detail (including its spectral properties and in stereo), determine the surface and exosphere compositions, measure the magnetic field and the interaction of high-energy particles with it, measure the topography to high precision, and determine the size of any fluid outer iron core. It will also determine the composition of the polar deposits, and of course determine the origin and age of the peculiar radar anomalies. Because it will determine the composition of the surface in detail we should be able to decide between the competing theories for Mercury's large iron core. This in turn will help to better understand the origin and evolution of the terrestrial planets.

The second planned Mercury mission is an ambitious European Space Agency "Cornerstone" mission that will consist of two orbiters and a lander. This mission is called BepiColombo after an eminent Italian celestial mechanician (Giuseppe Colombo) who died in 1984. Among other important things, he discovered that Mariner 10 could make additional encounters with Mercury instead of just one. Two additional flybys were accomplished and produced much more information about Mercury than would otherwise be possible, thanks to Colombo's work. Recognition at last. Finally, after 28 years of waiting, Mercury will be intensively explored. If all goes well we will have an in-depth look at Mercury by the end of this decade and the beginning of the next. It will no longer be the poorly known planet that has intrigued me for so long. I will finally get to see what those radar features really are, and to see the unknown side that has frustrated me for so long. Although most of my current questions will be answered, I am sure the new data will raise many more questions, but that is how science works. Now my goal is to live long enough to see all this happen. (I was 40 when Mariner 10 flew by Mercury, and I will be pushing 76 when MESSENGER goes into orbit.)

Further reading

Strom, R. G. (1987). *Mercury: The Elusive Planet, Smithsonian Library of the Solar System*, Smithsonian Institution Press.

Chapman, C. and M. Matthews (eds.) (1988). *Mercury, Vilas, F.*, University of Arizona Press.

Strom, R. G. (1998). Mercury, in *Encyclopedia of the Solar System*, Academic Press.

5

Return to the Moon!

Harrison H. Schmitt, University of Wisconsin

Harrison Hagan Schmitt, a native of Silver City, NM, is truly a Renaissance man: a geologist, pilot, astronaut, administrator, businessman, writer, professor, and US Senator from New Mexico. He received his B.Sc. in Science from Caltech, studied as a Fulbright Scholar at Oslo, and attended graduate school in geology at Harvard. "Jack," as he is colloquially known, was selected to be an astronaut in 1965, as a member of the first-ever class of scientist-astronauts. This of course, led to his journey to the Moon aboard Apollo 17 in late 1972. I met Jack in 1999 at a conference on the Moon, when he corrected me about the placement of a mass spectrometer unit deployed on Apollo 17 during a talk (to over 300 scientists). When I asked how the voice in the back of the room was so sure he knew so much, he simply replied, "Because I was there."

Create commercial enterprises based on resources from space that, taken as a whole, support the preservation of the human species, its freedom, and its home planet.

Current vision statement of the Interlune-Intermars Initiative, Inc.

Trips to the Moon figured prominently in the history of the world during the latter half of the twentieth century. A return to the Moon in the first decades of the twenty-first century may be even more significant. My interest in this possibility stems from having participated in the exploration of the Moon's Taurus-Littrow Valley as the only scientist to go to the Moon and the Lunar Module Pilot on the last of the Apollo Missions, Apollo 17. This personal opportunity came as a result of President John F. Kennedy's 1961 challenge to Americans, "to go to the Moon and return safely to Earth." Kennedy's inspiration coincided with a remarkable superposition of four social phenomena: public concern about the future, a sufficient base of technology, a catalytic and focusing event, and a leader who recognized a unique opportunity. The coincidence of these phenomena in the America of the 1960s provided the foundation for the success of Apollo as it did earlier for Thomas

Jefferson's Louis and Clark Expedition, Theodore Roosevelt's Panama Canal, and other critical endeavors in the history of the United States.

Not all great undertakings are assured of success, however. Apollo 11 succeeded in landing on the Moon on July 20, 1969, because competitive bidding brought the best of industry to the job, conservative engineering established strong margins of performance and safety, and highly flexible but disciplined management kept the ultimate objectives in perspective. Most importantly, nearly 500 000 highly motivated men and women, mostly in their mid twenties, believed that meeting President Kennedy's challenge was the most important contribution they could make with their lives. Ten years of 16-hour days, seven-day weeks, and the inevitable wear and tear on families could not be sustained without such a belief. No amount of money would have bought us the quality control, attention to detail, teamwork, and spur of the moment innovation that became the hallmarks of Apollo.

As President Kennedy appears to have anticipated, Apollo's success contributed in profound ways to the successful conclusion of the Cold War. Émigré reports and post-Cold War examination of Soviet records indicate that Apollo created a belief in the minds of the leadership of the Soviet Union that President Ronald Reagan's 1983 Strategic Defense Initiative probably would be successful as well, ultimately leading to a break-up of Soviet communism. Apollo also established for all human beings a new evolutionary status in the Solar System. As a consequence, young people alive today realistically can think about living in settlements on the Moon and Mars. They can anticipate helping their home planet survive, as America helped former homelands in Europe and Asia defeat oppression in the twentieth Century. All in all, we have had an unprecedented and continuing return on the 1960s investment in a "race to the Moon." Both Americans and Russians can be proud of the eventual results of their competition in that race.

Harvest Moon

Assistance to the Earth from future settlers of the Moon will come as a direct result of the scientific discoveries of Apollo, echoing earlier events in the history of the United States. The explorations of President

Thomas Jefferson's 1803 Louisiana Purchase by Meriwether Lewis and William Clark lay the foundations for the growth of the economy and power of the United States. Theodore Roosevelt's project to construct a canal and lock system across Panama made the United States a naval power on two oceans and produced an explosion in medical, construction, and electrical technology. Similarly, the exploration of the Moon by Apollo astronauts created the Earth's first preeminent space-faring nation and stimulated rapid advances in most fields of engineering. Additionally, Apollo laid the foundations for future terrestrial energy alternatives to fossil fuels, the growth of a lunar economy, and the settlement of the Solar System by humans. In anticipation of such implications, and even before the first lunar landing, Apollo Spacecraft Project Manager George Low and others had introduced lunar science as the second major objective of Apollo. Five other scientists and I were selected as the first scientist astronauts because of that foresight. Plans were made and contracts issued for special equipment, landing site selection, and extended missions. Astronaut training began to emphasize the knowledge and skills necessary to conduct meaningful fieldwork on the Moon in addition to operational mission training.

As a consequence of Apollo lunar exploration by 12 Americans and the more recent robotic exploration and scientific analysis built on that foundation, we have detailed first- and second-order understandings of the nature and history of the Moon, the smallest of the terrestrial planets. By extrapolation, we gained vastly improved insights about the history of other terrestrial planets – Earth, Venus, Mars, and Mercury. We have scientific visibility into the first one and a half billion years of the geologically clouded history of the Earth, including the origin of life, not accessible by any other means.

For example, it is now clear that the origins of the Moon and our home planet were closely related. After its accretion about 4.6 billion years ago, an ocean of hot magma existed on the Moon for about 50 million years during which mineral crystallization and density separation in that ocean caused the differentiation of a crust and a mantle. For the next 700 million years, intense bombardment by asteroids and/or comets cratered and pulverized the lunar surface and the surfaces of other terrestrial planets, ending about 3.8 billion years ago. Great surface eruptions of lava then dominated the next billion years, gradually dying out as the Moon cooled. Many other details of lunar history are known, but it may be most interesting to note the match between the appearance of isotopic evidence of life on the water-rich Earth with the end of the great bombardment, both occurring 3.8 billion years ago. The end of the extraordinary impact violence at this point in Solar System evolution may have finally permitted simple, replicating life to form at the surface of the Earth from a clay soup rich in water, complex organic molecules, and other necessary components. The same process may have begun on Mars only to be arrested later by the loss of its oceans and atmosphere.

Bucks for Buck Rogers

Most importantly, for future inhabitants of Earth and space, we know from the Apollo lunar samples that fusion energy resources (solar wind Helium-3, a light isotope of normal Helium) exist in the pulverized upper several meters of the Moon's surface. (Fusion energy is produced by combining atoms, whereas fission energy is produced by splitting atoms.) These potentially commercial energy resources provide both a long-term alternative to our use of fossil fuels on Earth as well as the basis for future lunar and Martian settlement. Further, by-products of the extraction of Helium-3 from the lunar surface can sustain the future travelers and settlers of deep space with water, oxygen, hydrogen fuel, and food. I doubt, however, that the United States or any government will initiate or finance a return of humans to the Moon or a human expedition to Mars in the foreseeable future. Governments, particularly that of the United States, do have a very pragmatic excuse for turning their financial backs on the potential of a return to the Moon. I learned during a term as a US Senator that it is very difficult, politically, to commit to the long-term allocation of the required taxpayer-provided funds for space or any other so-called "discretionary" activities. This would be true under any circumstances but is made impossible now by the inability of governments to fund retirement and health security for the elderly and the poor by means other than income transfer from one generation to the next. Income transfer will lead to higher and higher tax rates on the children of the World War II Baby Boom as their parents begin to retire early in the twenty-first century.

The private sector, on the other hand, may find a business rationale for a return to the Moon, based on the economic value in the extraction of lunar Helium-3 and its use as a fusion fuel on Earth, providing a future economic and environmental alternative to fossil fuels. This possibility has been studied extensively by my colleagues at the University of Wisconsin-Madison and myself since 1985 and continues to appear to be a feasible approach to a return to the Moon and to providing clean terrestrial energy in the future. In addition, by-products of Helium-3 extraction from the pulverized lunar surface soils will include hydrogen, oxygen, and water – valuable materials needed for consumption by humans elsewhere in space. Thus, the next return to the Moon probably will approach work on the lunar surface very pragmatically with humans in the roles of exploration geologist, mining geologist/engineer, heavy equipment operator/engineer, heavy equipment/robotic maintenance engineer, mine manager, and the like. To be successful, of course, a lunar resource and terrestrial fusion power business must be based on competitive rates of return to investors, innovative management of financial and technical risk, and reasonable regulatory and treaty oversight by government.

The long-term business case for private sector involvement in a return to the Moon most directly relates to terrestrial needs for clean energy. The global demand for energy will likely increase by a factor of eight or more by 2050. This will be due to a combination of needs reflecting the doubling of world population, new energy-intensive technologies, demands to avoid the adverse consequences of climate change, and aspirations for improved standards of living in the less-developed world. Lunar Helium-3, with a resource base in the titanium-rich basaltic soils of Mare Tranquillitatis of at least 10,000 tons, represents one of several potential energy sources to meet this rapidly escalating demand. The results from the 1997–99 Lunar Prospector orbiting neutron spectrometer analyses suggest that Helium-3 also may be concentrated at the lunar poles along with solar wind hydrogen.

The energy equivalent value of Helium-3 delivered to future fusion power plants on Earth would be about $3 billion per ton relative to

$21 per barrel crude oil. The domestic US electrical power market is worth approximately $120 billion, annually. Some 40 tons of Helium-3 contains enough energy to supply that market's needs for one year. These numbers illustrate the theoretical magnitude of the potential business opportunity in a return to the Moon. In addition, the technology and facilities required for success of a lunar commercial enterprise will make possible and reduce the cost of continued scientific investigations on and from the Moon; space station re-supply; exploration and settlement of Mars; asteroid interception and diversion to prevent impact on the Earth; and many other future space activities.

Mining, extraction, processing, and transportation of Helium-3 to Earth, and the use of by-products in space, requires new innovations in robotic and long-life engineering but no known new engineering concepts (Figure 5.1, see color plate section). A business enterprise based on lunar resources will be driven by cost considerations to minimize the number of humans required for the extraction of each unit of resource. Humans will be required on the Moon, on the other hand, to reduce the business risk of lunar operations; to prevent costly breakdowns of semi-robotic mining, processing, and delivery systems; to provide manual back-up to robotic or tele-robotic operation; and to support other lunar activities in general. The creation of capabilities to support mining operations also will provide the opportunity for renewed scientific exploration at much-reduced expense, with the cost of capital for launch and basic operations being carried by the business enterprise. During the early years of operations the number of personnel at a lunar resource extraction settlement will be about six per mining/processing unit plus four support personnel per three mining/processing units. Cost considerations also will drive business to encourage or require personnel to become settlers, provide all medical care and recreation, and provide technical control of most or all operations on the Moon.

The questions probably will always be asked: "Why humans in space or on the Moon? Wouldn't robots be better and safer?" Setting aside the inherent desire for human beings to "be there'" where ever "there" may be, I know from personal experience that on the Moon, humans contribute to space operations in unique and valuable ways. They will provide instantaneous observation, interpretation, and assimilation of the environment in which they work and a creative reaction to that environment. Human eyes, experience, judgment, ingenuity, and manipulative capabilities are unique in and of themselves and highly additive in synergistic and spontaneous interaction with instruments and robotic systems. Due to inherent communication delays and the cost of returning samples and providing mission support, the deeper into space human beings desire to go, the more important will become these unique human attributes.

Near term, the more critical and enabling question is: "Can you cut the cost of access to deep space from the approximately $70,000 per kilogram (including the theoretical cost of private capital) required by the Apollo Saturn V rocket and by existing heavy lift technology?" Heavy lift launch costs constitute the largest cost uncertainty facing initial business planning. However, many factors, particularly long-term production contracts, promise to lower these costs into the range of $1,000–$2,000 per kilogram. Also contributing to a reduction in launch costs will be nearly 40 years engineering experience with heavy lift rockets, many new technologies that have never been applied to such rockets, and a clear focus on a set of business and financial requirements.

Another critical question relates to the technology base for the use of Helium-3 as a fusion fuel. Inertial electrostatic confinement (IEC) fusion technology (Figure 5.2) appears to be the most attractive and least capital-intensive approach to terrestrial fusion power plants. Although great amounts of public funds have been spent on fusion research over the last half-century, this research has almost exclusively focused on the technology of non-electrostatic confinement devices. These technologies suffer from numerous disadvantages in their possible application to commercial electrical power plants, including very high capital and operating costs, large minimum operating size, relatively low conversion efficiency through a heat cycle, and radioactive fuel and waste products. In contrast, IEC technology inherently offers the potential for low capital costs, size flexibility, high conversion efficiency through direct conversion of charged particles, and non-radioactive fuel and no

Figure 5.2

D/H3-3 fusion in an Inertial Electrostatic Confinement device for laboratory experimentation. Courtesy of the Fusion Technology Institute, University of Wisconsin-Madison.

radioactive waste. Over the last two decades, steady progress in the advancement of IEC fusion technology has been made by my colleagues at the University of Wisconsin-Madison's Fusion Technology Institute under the guidance of Professor G.L. Kulcinski.

A private enterprise approach to developing lunar Helium-3 and terrestrial IEC fusion power would be the most expeditious means of realizing this unique opportunity. In spite of the large, long-term potential return on investment, access to capital markets for a lunar Helium-3 and terrestrial fusion power business will require a near-term return on investment, based on early applications of IEC fusion technology. The most obvious such application will be in the low-cost, point-of-use production of short half-life medical isotopes.

The international space treaty environment forms an important backdrop to a return to the Moon for its resources. The only space treaty related to the use of space resources to which the United States is a party is the 1967 "Treaty on Principles Governing the Activities of States in the Exploration and Use of Outer Space," or Outer Space Treaty. The Outer Space Treaty specifically provides a generally recognized legal framework for such use. The Treaty does not contain specific rules relative to the extraction and use of lunar resources, however its provisions imply certain guidelines for such activities. Compliance with these guidelines by a legal corporate entity under the laws of the United States would be straightforward. The 1979 Moon Agreement (commonly referred to as the Moon Treaty) has confused the treaty environment somewhat, but that Agreement has not been ratified by major space-faring nations. If it were so ratified, it would build in a high degree of uncertainty that is antithetical to private commercial activities on the Moon. The Agreement would, in effect, create a moratorium on such activities. The Agreement's mandated international regime would both complicate private commercial efforts and give other countries political control over the permissibility, timing, and management of all sanctioned commercial activities.

Return!

Thus, a return to the Moon inherently has both great potential and great challenges. The twentieth century, however, saw the beginning of the movement of the human species into space, best symbolized by the astronaut's photographs of the crescent Earth rising over the lunar horizon (Figure 5.3, see color plate section). A return to the Moon to stay early in the twenty-first century will be the most direct means of continuing the migration begun out of Africa hundreds of thousands of years ago and of continuing to realize its benefits on Earth.

Further reading

Neal, V. (ed.) (1994). *Where Next Columbus?*, Smithsonian Institution Press.

Melberg, W. (1997). *Moon Missions*, Plymouth Press.

Bond, P. (1993). *Reaching for the Stars*, Cassell.

Canup, R. M. and K. Richter (eds.) (2000). *Origin of the Earth and Moon*, Arizona University Press.

Plate Section

Figure 2.1. View of the Grand Tetons an early influence. This is one of the American landscapes painted by my grandfather after he came to the US from Switzerland in 1892. His paintings hung in our home and in relatives' homes, which subliminally accustomed me to the idea of paintings as a means of expression. This view was painted in 1952 in an impressionist-influenced style, typical of many European artists who painted in the Western US in the late 1800s and early 1900s. A bullet hole was put through the piece during a hold-up in a coin shop, operated by my uncle, in Houston, adding some typical Texan Americana! I repaired it. Painting by Andrew Hartmann.

Figure 2.2. The explosion over Siberia in 1908. An ongoing project is to read the eyewitness reports collected by Russian scientists about the Siberian mystery, now attributed to an asteroid or comet exploding in the atmosphere. Witnesses did not know what they were seeing, and described the sky opening up and pouring forth fire, but armed with modern knowledge we can reconstruct the scene from their accounts. This view is from a trading station, about 60–70 kilometers from the blast, where a witness was knocked about 20 feet off a porch where he was sitting. I used similar forests in Flagstaff, Arizona, as a model and painted outdoors from life. Painting by William K. Hartmann.

Figure 2.3. Formation of the terrestrial planets. This painting shows some of the features postulated about the formative conditions of the planets. A thick disk of dust and gas surrounds the early Sun, similar to dust disks seen around other newborn stars. Planetary bodies of various sizes are forming of aggregation of the dust particles into larger and larger bodies. The foreground nucleus of the proto-Earth is beginning to accrete a gaseous atmosphere. Painting by William K. Hartmann, commissioned by German planetary scientist, Gunter Lugmair.

Figure 2.6. On the night side of a comet. A challenge in astronomical painting is to think up effects and scenes that have not been visualized before. Until the 1980s, few scientists thought much about the surfaces of comet nuclei as solid, geological bodies. This painting shows the view from the night side of a comet. In the sky, the shadow of the nucleus is cast through the coma, creating a fuzzy dark blob. Streamers of the tail converge close to the anti-sunward direction. This painting, to my knowledge, is the first to illustrate such effects. Painting by William K. Hartmann.

Figure 3.1. Still life with Antarctic meteorite.

Figure 4.1. An artist's rendition of Mercury's huge iron core with a mosaic of the incoming side of Mercury from Mariner 10. The core is about 75% of the diameter, and the outer part may still be molten.

Figure 5.1. Concept for a mobile mining/processing machine for the extraction of solar wind volatiles, including Helium-3, from lunar soils. Courtesy of the Fusion Technology Institute, University of Wisconsin-Madison.

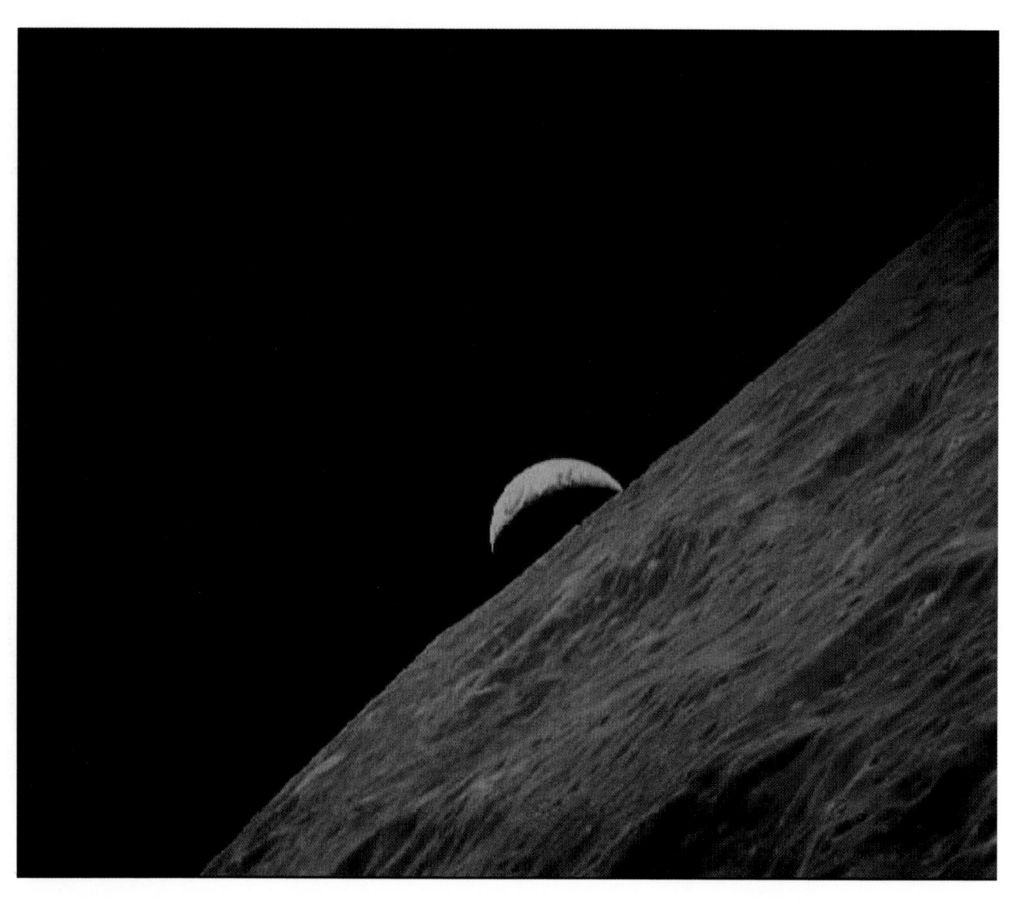

Figure 5.3. Apollo 17 photograph of the crescent Earth rising over the far-side lunar horizon. Courtesy of NASA.

Figure 6.1. A painting illustrating the formation of the Moon by the impact of a Mars-sized projectile on to the young Earth, by William K. Hartmann (with permission).

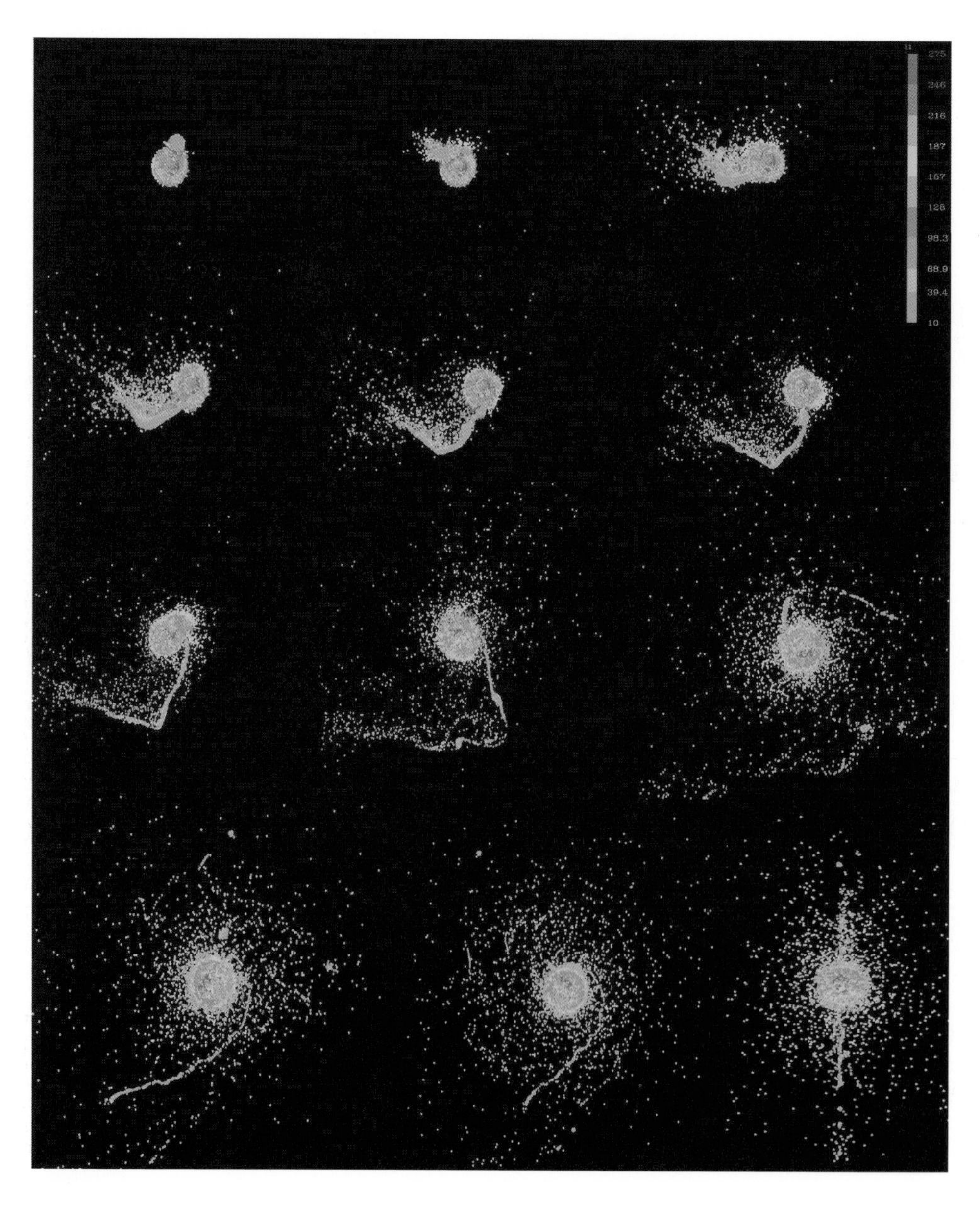

Figure 6.2. A time series of a simulation of a potential Moon-forming impact (from Canup and Asphaug 2001). Color scales with the degree of heating of the material, with blue to red corresponding to cool to highly heated material. In the first frame in the upper left, the Mars-sized protoplanet collides with the young Earth in a counter-clockwise sense, launching material into Earth-orbit and finally leaving a disk of hot debris. The final frame on the bottom right, shown after about 24 hours of simulated time, is the system viewed on-edge.

Figure 7.5. Color mosaic of the rover measuring the chemical composition of the rock named Moe on Mars. This image was used for the cover of *Science* (volume 278, pages 1734–1774, December 2, 1997) in which the scientific results of the Mars Pathfinder mission were first reported. Special issues of the *Journal of Geophysical Research, Planets* (volume 104, pages 8521–9096, April 25, 1999; and volume 105, pages 1719–1865, January 25, 2000) also featured the scientific results of the Mission.

Figure 8.4. Pluto map showing polar caps and dark equatorial region.

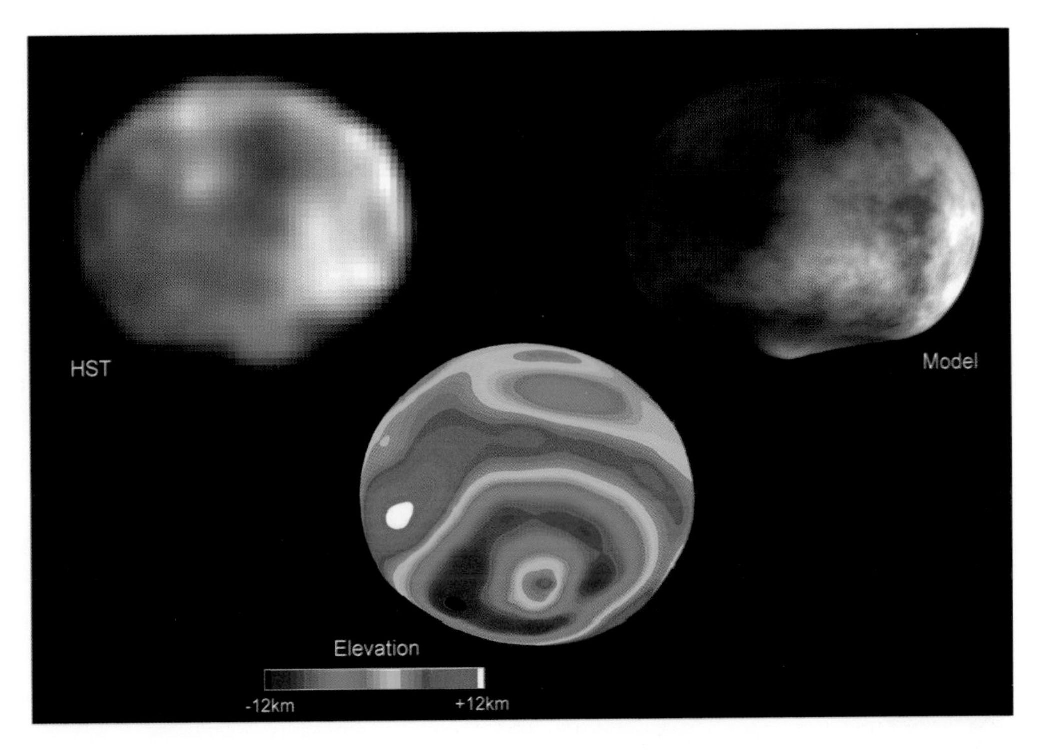

Figure 8.5. HST images and model of Vesta showing the huge impact basin.

Figure 9.2. Photograph of chamber used in the laboratory to make solid organic material from Titan-like mixtures.

6

Making moons

Robin Canup, Southwest Research Institute

Robin Canup grew up in New York and took her undergraduate training in physics at Duke. She went on to the graduate school in planetary science at the University of Colorado. I first met Robin when she was entering the University of Colorado; I was just leaving to post doc. Beyond being a consummate planetary scientist, Robin is also an accomplished ballerina and a wonderful departmental manager. She and her husband Rick Mihran live in the mountains outside Boulder, where they are renovating yet another home, for entertainment.

Are worlds like ours common or rare? Are we alone in the universe? These are questions that transcend the day-to-day, or even generation-to-generation happenings. Although not the kind of issues that would be typically characterized as having great practical consequence – they do not allow one to build a house, or to grow food, or to fix a car – they are the stuff that is in the back corner of the mind of every astronomer. They arise from some seed of wonder planted in each of us in our youth, perhaps long since replaced with more sophisticated and pragmatic pursuits, but still there nonetheless, providing that ongoing desire for the pursuit of the unknown, the quest for understanding. The quest for that kind of intangible is not just the realm of science of course; it is also at the heart of art, philosophy, and many other disciplines. In astronomy, this type of pursuit is often referred to as "the search for astronomical origins," that is, determining how the universe and the objects within it came to be.

From our particular vantage point in the universe, the closest planetary bodies are our planet and its Moon, and the origin of these two objects has been the central theme of much of my own scientific research. While astronomy is a broad field that includes many fascinating topics, it is the formation and history of the objects nearest to us that I have personally found compelling. Part of the appeal of studying the Earth and Moon is their inherent human connection: Earth is of course our home, and its Moon our companion throughout history.

Every human has gazed upon the Moon, and each day all living things reap the rewards of those particular series of events that left our planet in its wonderfully habitable form. I also enjoy this area of research because of its inherently interdisciplinary nature: the wealth of known geological, chemical, and physical properties of the Earth and Moon provide a rich multitude of constraints for any scientific model and modeler.

Earth and Moon

From dating of lunar and terrestrial materials, we believe that both the Earth and Moon are extremely ancient, having formed some 4.5 billion years ago, or 50–100 million years after the oldest known grains in the Solar System. To hypothesize about events that occurred so long ago – and which are no longer observable today – requires detective work, hunting through the detailed properties of current objects to find those that may hold clues to their early origins.

It has been appreciated for centuries that our Moon is quite unique. Our rocky neighbors either have no moons (Mercury and Venus), or, in the case of Mars, only very small, irregularly shaped ones. In contrast, the Moon is a magnificent planetary body in its own right – thousands of kilometers across (larger even than the planet Pluto), with an intricate surface marred with craters, rugged highlands, and plains filled with ancient volcanic lava flows. The Moon is under dense compared to the other rocky planets, with a bulk mass density that is 40% lower than that of the Earth. The only way to account for such a low density is for the Moon to have much less than its proportionate share of very high-density metals, such as iron and nickel. In the Earth, such elements compose a central dense core that makes up about 30% of its mass. But the Moon's core – if it has one at all – comprises no more than a few percent of its mass, an oddity that any viable theory for lunar origin must somehow accommodate.

Another clue concerning the Moon's beginnings is found in the current lunar orbit. The Moon currently orbits the Earth at a distance of 384,000 kilometers, or about 60 Earth radii. But it was not always so distant. The Moon's gravity acts to distort the figure (that is, the shape) of the Earth slightly, raising tides, most noticeably in the oceans. In the time it takes for a rising tide to form, a given point on the Earth's surface rotates ahead of the Moon, so that from the perspective of the Moon, the tidal distortions on the Earth form somewhat in front of it. The interaction between the Moon and the slightly non-spherical figure of the Earth slows the Earth's rotation and causes the Moon's orbit to expand outward.

Currently the expansion rate of the Moon's orbit is only a few centimeters per year, but this rate is a strong function of distance from the Earth, so that early in its history, the Moon was evolving away from the Earth by as much as 100 kilometers per year. Throughout this Earth–Moon gravitational interaction, a quantity known as angular momentum is conserved; this quantity is a measure of the rotational motion contained in both the Earth's spin and the Moon's orbit. From the Earth's current 24-hour day and the Moon's present orbit, one can thus extrapolate back in time and predict what the initial length of a terrestrial day was when the Moon was young and close to the Earth. Four and one-half billion years ago, the newly formed Earth had a day of only about five hours – a very rapid rotation compared to the current rotations of other planets in our Solar System. At that time, the Moon is predicted to have been in close proximity to the Earth, about 15 times closer than it is currently.

Prior to the 1970s, there were three main theories for lunar origin: fission, capture, and co-formation. The fission hypothesis imagined that early Earth had spun so rapidly that it became rotationally unstable, finally relieving its internal stresses by cleaving a large chunk from its equator that then became the Moon. Capture proposed that the Moon was an intact planet, forming in the same general area of the Solar System as the Earth, that passed so close to the Earth that it became gravitationally trapped in an Earth-bound orbit. Lastly, co-formation viewed the Earth and Moon as siblings that had grown alongside one another as an ancient bound pair. These three competing theories had vied for dominance for some time, each with its own distinct implications for the history and origin of the Earth. Indeed, a particular scientific objective of the Apollo lunar exploration program had been to determine which of these theories for lunar origin was correct.

However, new information often brings new and unexpected constraints, which sometimes complicate rather than simplify pre-existing concepts. And indeed, as the lunar samples collected during the Apollo program were analyzed, they revealed more lunar intricacies. In some respects, lunar rock chemically looked surprisingly similar to the Earth's mantle. In particular, the pattern of abundances of the isotopes of oxygen in terrestrial and lunar samples was indistinguishable, and quite distinct from those seen in meteorites believed to have originated in the asteroid belt. This seemed to suggest that the Earth and Moon formed from a common supply of material. However the lunar rocks differed greatly from their terrestrial counterparts in a key respect: they were lacking in certain elements (known as volatiles) that are easily vaporized when a substance is heated. It appeared that the material that formed the Moon had been heated to a greater degree than that which formed the bulk Earth, leaving the current Moon dry and comparatively lacking in elements like water and potassium.

In 1975–1976, two groups, working from two quite different perspectives, proposed a new idea for lunar origin. Alastair Cameron (then at Harvard, now at the University of Arizona) and a young post-doctoral researcher, William Ward (also then at Harvard), realized that the rapid initial spin of the Earth could have been the result of a single large impact, if the collision was nearly grazing and involved an impacting body approximately as large as the planet Mars. They proposed that vapor produced during the extensive impact-induced heating could assist in lofting some of the collisional debris into orbit around the Earth, where it might then accumulate into the Moon. At approximately the same time, William Hartmann and Donald Davis (Planetary Science Institute) were independently developing models of planet formation, and observed in their simulations that collisions between planetary-sized objects were common, leading them to suggest that one such collision or sequence of collisions with the young Earth might have ejected material into Earth orbit, from which the Moon might then have formed. Both groups recognized that an energetic collision might naturally cause its remnants to be severely heated, thus potentially accounting for why the lunar rocks appeared so depleted in volatile elements. In addition, they realized that the material ejected into orbit around the Earth might preferentially originate from the low-density outer mantles of the colliding objects, with little or none originating from the iron cores, thus potentially explaining why the Moon is oddly depleted in iron. The "giant impact theory," as it was later to be known, was born.

In 1975, I had no knowledge of the giant impact theory: I was seven years old. My closest experience with planetary science involved my father reading to me from my favorite book, entitled *The Earth for Sam* by W. Maxwell Reed.

Decades later, a senior colleague that I would then know as Bill Ward would tell me of the many years of skepticism that followed the original proposal of the impact theory. The idea that a Mars-sized planet had collided with early Earth was seen as radical, largely because a pseudo-scientific author by the name of Velikovsky had received wide public attention for his ideas about planets wildly colliding with one another in modern times, thus providing a highly controversial means of accounting for various biblical events. Self-respecting planetary scientists of the day were quite wary of such notions. At the time, the standard idea for how planets like the Earth formed involved the ongoing sweep-up of small material orbiting nearby to the growing planet. Requiring that one such collision with the Earth involved an object as large as Mars seemed not only improbable, but also *ad hoc*.

Aloha Moon

In December of 1984, a scientific meeting on the topic of the origin of the Moon was held in Kona, Hawaii, at which about 60 experts debated the strengths and weaknesses of all of the various theories for lunar formation. They even went so far as to create a quasi "report card," in which every origin theory was graded on its ability to account for each of the main characteristics of the Earth–Moon system. One by one, theories other than the giant impact hypothesis were weakened by accumulating implausibilities. New simulations had shown that fission required too much angular momentum; capture could not easily account for the Moon's lack of iron, and seemed too dynamically unlikely. Co-formation produced a Moon that was iron-rich like the

Earth unless some rather complex filtering mechanisms were invoked, and even then it was difficult to explain the high angular momentum of the Earth–Moon system.

In contrast, the impact theory seemed potentially able to account for all of the main characteristics of the Earth–Moon system. This once "radical" theory also gained credibility from new models of planet formation, which suggested large impacts might be common events in the end stages of terrestrial planet formation. Such models demonstrated that the relatively quiescent stage of planetary growth continued only until young planets grew to sizes ranging from lunar to Mars-sized, and that the final stages were characterized by collisions among tens to hundreds of these large, planet-sized bodies. In the course of the many impacts apparently required to yield the final four terrestrial planets, it did not then seem so unreasonable that one of the impacts would be of the type required to yield our Moon. By what was in some sense a process of elimination, the giant impact theory emerged from this conference as the leading contender for lunar origin, somewhat to the surprise of even the meeting organizers I am told.

At the end of 1984, as the impact theory was beginning to take hold in Hawaii, I was a third of the way around the globe in New York, a junior in high school consumed by dancing Sugarplum Fairy in "The Nutcracker." I was an aspiring ballerina; I loved the combination of creative expression and demanding technique. But I was also a good student, particularly in math and science, and I was beginning to feel that I needed to decide between pursuing ballet as a career and going to college. I worried that the life of a professional dancer appeared to be one of struggle, with an agonizingly short career life. I also started to recognize the potential frustration of working in a profession where a single physical feature was often used as the primary metric of worth. I had attended an audition with a hundred other girls, which began with our being lined up in single rows. The director had walked around the room, looked at the shapes of our legs and feet, and then dismissed all but about ten girls. I had not been one of those ten. That spring, I applied to colleges, with plans to major in mathematics and biology and then go on to medical school.

But plans one makes in life often evolve in ways you can not anticipate. After one year of college chemistry at Duke University, I realized that the "pre-med" culture did not suit me. A wonderful professor in introductory physics – Alec Schramm – inspired me to switch my major to physics (much to the delight of my physicist father). To my surprise, out of 1,600 students in my class at Duke, there were only 16 physics majors, and the classes were accordingly small and intensive. My favorite upper-level physics course turned out to be a one-semester elective class in astrophysics, taught by an extraordinary teacher, John Kolena, who also teaches high school students at the nearby North Carolina School of Science and Math. With the job market in 1990 slowing as I neared graduation, I made plans to attend graduate school in astronomy.

By 1993, I was a graduate student at the University of Colorado studying the dynamics of planetary ring systems. By that time, the giant impact theory for lunar origin enjoyed wide support – it was the standard in every major astronomy textbook – and many researchers that had once worked on lunar-origin models had moved on to other topics. But it was that year that I started thinking about the origin of the Moon for the first time, all because of a single conversation with one of the theory's originators, Bill Hartmann.

We were at a scientific conference, and I had just given my first scientific talk on simulations I had developed for describing how numerous small satellites, near the outer portions of planetary ring systems, might have originated. After my talk, Hartmann approached me and asked, "Have you ever thought of applying your models to the Moon? I keep trying to talk someone into modeling how the Moon would actually have accumulated after the giant impact." We spoke for a bit, and modeling the Moon's formation did sound like a very interesting problem to me. However, I kept thinking that surely this was a problem that someone had already worked out. It is of course the nightmare of every graduate student to work on a problem for years, only to discover that it has already been solved in a previous study. To get a Ph.D., one has to make a new contribution to his or her field, and, from the perspective of a graduate student, it often seems as if all of the good problems have already been worked. As one advances in the field, your perspective shifts, and you are often amazed at how many important problems are still yet to be tackled. But then I was still a graduate student, and my thesis advisor, Larry Esposito, wisely advised me to research all of the work that had been done on the impact theory before deciding to work on it as part of my thesis.

Moon rise

In the decade following the Kona meeting, a new and exciting computational technique had been developed that allowed researchers to simulate the type of impacts that might yield the Moon. This mathematical technique, known as "smooth particle hydrodynamics," or SPH for short, described the colliding planets as a large number of individual spherical particles that each interacted gravitationally, and whose individual thermodynamic behaviors (for example, heating/cooling/ phase changes) were tracked as a function of time. The "particles" in these types of simulations have no true physical meaning; they are simply a mathematical device to simplify the behavior of a highly complex physical system into a number of representative points, small enough to be tracked in a computer simulation.

There had been a general concern prior to the Kona meeting that it might be impossible for an impact to eject sufficient amounts of material into the Earth's orbit to yield the Moon. The gravitational force of a spherical planet is such that an object launched from its surface will either re-impact the planet, or escape from the planet's gravitational well altogether. But to form the Moon, a few percent of the Earth's mass would have to be ejected into trajectories that would orbit the Earth well above its surface. Was this feasible?

The first simulations of potential lunar forming impact events seemed to suggest that indeed it was. Impact simulations performed by Willy Benz (then at Harvard, now at the University of Bern, Switzerland), Al Cameron, and others found that off-center collisions with the Earth seemed the most effective at lofting material into orbit; interestingly, these were also the type of collisions necessary to account for the Earth's rapid early spin rate, which was a compelling new factor in favor of the impact theory. A head-on collision would not have provided the Earth with any spin, while a grazing impact by a Mars-size or larger object could spin up the Earth to an appropriately short rotational day.

By 1991, the new SPH simulations had demonstrated that a giant impact could place a lunar mass or so worth of material into orbit, and that the impact parameters could be adjusted so that the ejected material contained little or no iron. Figure 6.2 (see color plate section) is a modern-day example of an SPH simulation of a Moon-forming impact, which shares many commonalities with the earlier models. Initially, the impact event greatly distorts target Earth, which reassumes an oblate spheroid shape by the end of the simulation. Most of the impactor accumulates on to the Earth while some portion of its mantle ends up in orbit, together with some material from the proto-earth's mantle. The impact energy is so large that it is likely sufficient to melt the Earth completely, and possibly vaporize some of its outer portions. The entire impact event – from the initial collision to the resettling of the Earth – occurs in less than a day of real time.

Given the results of such simulations, enthusiasm for an impact origin of the Moon grew quickly, as did the general concept that large impacts might have been responsible for a wide variety of planetary characteristics. It was proposed that Mercury's high density – a result of this planet having an atypically large iron core – arose when that planet was hit by a large, head-on impact which caused its outer mantle material to be stripped away, leaving its original large core intact within a smaller planetary remnant. The odd planet–satellite pair of Pluto and its moon Charon resembled the Earth–Moon system in some respects; that is, Charon is very massive compared to Pluto, and the system as a whole contains a large amount of angular momentum. An impact origin scenario for Pluto–Charon thus soon became favored as well. Then there was Uranus, that oddly tipped planet whose north pole is angled by nearly 90 degrees with respect to the normal to its orbit plane (the Earth's rotational pole tilt angle, known as its obliquity, is only 23.5 degrees by comparison). Had Uranus been knocked on its side by a large impact event? The paradigm of "giant impacts" had arrived, and suddenly what had seemed outlandish to some a decade earlier was now mainstream. Our thinking of the early Solar System as a plodding and predictable place gave way to the notion of planet-sized objects careening into one another in wild, stochastic ways. Just a little perturbation in the initial planetary configurations, and the Moon-forming impact might not have happened.

To top it all off, having the Moon – as a result of this now-believed chance planetary encounter – began to appear to be critically tied to the Earth's climate and ability to support life. The broad seasonal climatic trends on our planet are a sensitive function of our planet's obliquity. With our Moon, the variation in the Earth's obliquity is only a degree or so, and even this may be a significant enough change to, for example, effect ice ages. But in 1974, Bill Ward studied what the behavior of Earth's rotational axial tilt would be *without* the Moon. He discovered that amazingly, in the absence of the Moon, perturbations from the other planets would cause the obliquity of the Earth to vary widely between very high and very small values. Later numerical simulations by Jacques Laskar (of CNRS, in Paris) would show that the terrestrial obliquity without the Moon could get as large as 85 degrees, with chaotic variations on time scales of approximately millions of years. Imagine if the Earth's axis variation caused a global freezing of the oceans every few million years, interspersed with periods of summer–winter climate extremes. Even a moon less than half as massive as our own would not have been sufficient to keep the Earth from experiencing significant obliquity variations. Add to this the role of the Moon in inducing daily ocean tides (believed important for the evolution of land-dwelling life) and a strong connection is made between our Moon and our habitable Earth. With no Moon – or even a smaller moon – Earth may not have been the hospitable world we know today.

Moon making

Given the widespread enthusiasm for the giant impact theory, as well as the important effects of the Moon on the Earth, it came as a great surprise to me back in 1993 to discover that models for the actual accumulation of the Moon from the impact-ejected debris had never really been developed, and so this topic dominated the latter half of my dissertation and my early post-doctoral studies.

In the first part of my thesis work, I had developed a model to describe how small satellites accumulate in the outer portions of a planetary ring system. A planetary ring is composed of an immense number of individual particles, each on their own orbit around the planet. The ring particle orbits are so close to the planet that the planet's gravity is very effective at pulling apart particles in momentary contact with one another; this shearing ability is known as the tidal force. A planet's tidal force is particularly important within a radial distance from the planet known as the Roche limit, defined as the closest distance (typically about 2–3 planetary radii) a fluid object with no internal strength can orbit without being torn apart by the planet's gravity. Planetary rings are generally located within their respective planet's Roche limit, and so as planetary ring particles collide, they are generally unable to remain gravitationally bound to one another. This is what keeps a planetary ring like Saturn's dispersed, preventing it from accumulating over time into a few large, discrete satellites, such as those observed at greater orbital distances.

In the case of the Moon, the SPH impact simulations performed by Cameron, Benz, and others had predicted that the majority of the material that was ejected into Earth orbit, following a potential lunar forming impact, formed a cloud of gas and debris, roughly centered in the Earth's equatorial plane, with an outer edge of only a few planetary radii. So while the Moon today is far from the Earth, it had its origins from a debris disk that was quite centrally condensed, with about half of the material interior to and about half exterior to the Earth's Roche limit. Thus the early protolunar disk was approximately the same scaled distance away from the Earth as Saturn's rings are from it.

The year 1995 was a busy one for me. I defended my Ph.D. thesis, and several months later, married my dear companion-in-life, Richard Mihran. That year, in the final portion of my thesis work, I also had developed (in conjunction with Larry Esposito) the first model for the accumulation of the Moon from impact-ejected debris which accounted for the effects of the orbiting material lying near and within the Roche limit. This was first accomplished by using a statistical model, and later in collaboration with others by using what is known as an "*N*-body simulation." In an *N*-body simulation, the motion of some ("*N*") number of objects, is tracked by explicitly calculating the gravitational force on each object caused by every other object in the simulation at each time step. Figure 6.3 is a graph showing the results of one such lunar accumulation simulation. When orbiting particles collide with low enough impact energies, the result is a gravitationally bound aggregate when the collision occurs outside the Roche limit; inside the Roche limit, most collisions result in rebounds rather than mergers. Collisionally produced aggregates continue to grow in size as they collide and sweep-up increasing numbers of other objects, forming progressively larger and larger bodies in a process known as accretion.

Based on this model, a first question to be addressed was why a swarm of debris orbiting close to the Earth would yield a single large moon, when we find systems of multiple moons and rings around the gas giant planets. Interestingly, we discovered that the large mass of the Moon played a key role in its final singularity: once a "moonlet" forming exterior to the Roche limit began to acquire a mass approaching that of the Moon, its gravitational influence on other orbiting material became so large that remaining material inside the Roche limit was perturbed on to the Earth. In cases where more than one moonlet accreted initially, the multiple moonlet system was unstable as it evolved, producing either collisions among the moonlets or a moonlet–Earth collision. All cases seemed destined to yield a single, rather than multiple, moon(s).

Our models also predicted that the Moon accumulated on an incredibly short time scale after the impact event. Near the Roche limit, an object completes one orbit around the Earth in only about eight hours; solid material outside the Roche limit can then accrete into a single Moon in only a few months. The orbiting material is likely too hot to be completely solid or even molten, with some portion of it potentially heated to the point of vaporization, and much may be within the Roche limit initially. In this case, a more prolonged period for the Moon's formation would be needed to allow for the disk material to evolve and cool. Even so, the entire time frame for lunar formation is predicted to be less than a century – incredibly short in the context of an overall planet formation process that spans on the order of 100 million years. A hundred years after the giant impact, the young,

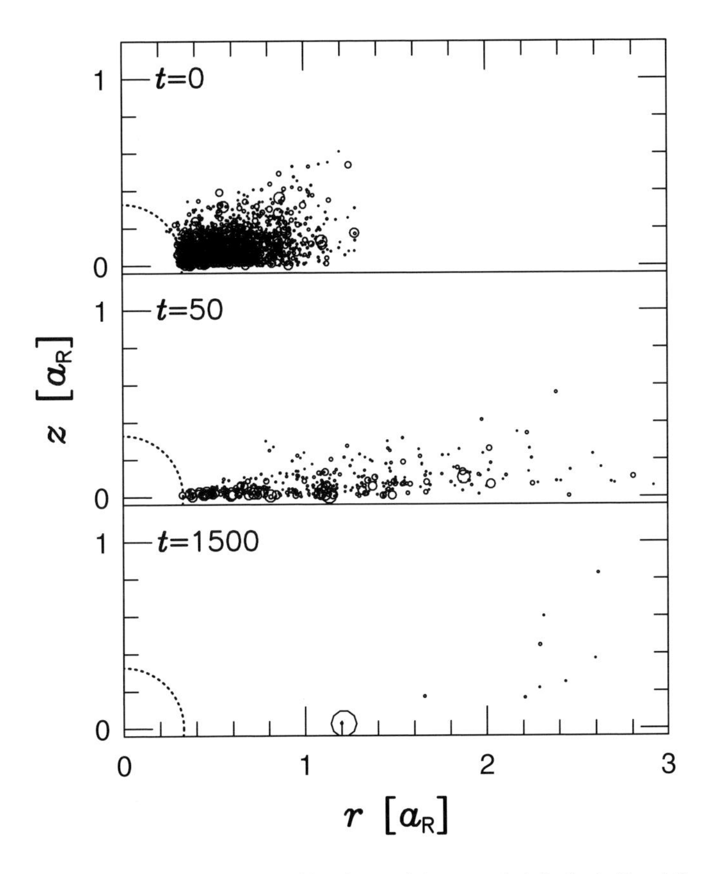

Figure 6.3

The impact-ejected debris orbiting the Earth (quarter circle in the bottom left corner), viewed on edge. Distances are shown in units of the Earth's Roche limit (a_R). As orbiting particles collide, they accumulate into larger and larger objects. After a few months to a year, a single large moon has formed just outside the Roche limit. From Ida, Canup, and Stewart, 1997.

molten Moon would have been orbiting at a distance of only about 25,000 kilometers from the center of the Earth – about 15 times closer than it is today. What an impressive sight in the sky it would have been, although the then molten Earth would not have provided a very hospitable viewing site!

Inconstant Moon

While the models of lunar accumulation appeared to explain why we have only one moon, they also showed that the accretion process, as we understand it, is very inefficient, with often less than half of the initially orbiting debris incorporated into the final moon. This, in turn, meant that a more massive initial debris cloud – one containing at least two lunar masses, or a whole lunar mass of material outside the Roche limit initially – would have been needed to yield a lunar-sized moon.

In the late 1990s, I began to collaborate with Al Cameron to determine which of his many giant impact simulations were the best Moon-forming candidates. Determining the type of impact necessary to form the Moon is important because it has broad ramifications for the physical (chemical and geological) histories of the Earth and Moon, and it allows for specific comparisons between predictions from the impact simulations and observed physical properties of terrestrial and lunar materials.

For several years, Cameron and I could identify only two categories of impacts that placed sufficient iron-free material into orbit to yield the Moon. One class involved an impactor with three times the mass of Mars, and an impact with more than twice the current Earth–Moon system angular momentum. Such impacts yield a Moon of the correct size, but leave the Earth spinning too rapidly, with an initial day of only about 2.5 hours instead of the approximately five-hour day needed to account for our current 24-hour day. Since the angular momentum of the Earth–Moon system has been very nearly conserved over the age of the Solar System, (probably less than 10% has been lost to interaction with the Sun) these high angular momentum impacts require the invocation of some mechanism to significantly slow the Earth's spin after the Moon-forming event, such as perhaps a second massive impact. Not exactly a simplifying result.

Another class of impacts that produce an appropriately sized moon yields the correct total Earth–Moon system angular momentum, but involves an Earth that is only 60% of its current mass. This "Early Earth" scenario implies that the Moon-forming impact occurred early in the Earth's accretion history. However, this model also has difficulties: if the Earth continued to accumulate solar-orbiting material in large amounts after the Moon formed, it would have been difficult to prevent the Moon from becoming contaminated with iron and volatile-rich material delivered by such collisions as well. Thus, this impact scenario shares some of the same drawbacks as the "co-formation" theory.

For several years I was beginning to wonder whether we had stumbled upon a major problem with the impact theory, as the above impact scenarios seemed to be much more restrictive and less elegant that the original impact hypothesis. An impactor *three* times the mass of Mars? Multiple giant impacts? Requiring the Moon to avoid accumulating material containing typical proportions of iron during the final half of Earth's accretion? Of course such circumstances cannot be ruled out *a priori*, but one becomes more concerned with a theory's merit as the number and combination of elements required for its success grows. The impact theory seemed once again, at least to me, somewhat *ad hoc*.

About two years ago, we began taking a somewhat different approach. Since the individual impact simulations are very computationally time consuming, with each one taking as long as several months to complete on a dedicated fast workstation, we started to investigate whether the simulations performed by Cameron displayed common trends when viewed in terms of generalized impact variables. Such a trend might relate, for example, how the amount of mass ejected changed as the impact angle or the size of the impactor was varied. If such trends did exist, they could then be extended to other types of impacts to predict what the outcome of a given impact would be, with the goal of identifying the most promising impact candidate without having to simulate every possible one. This proved to be a successful pursuit, as we (myself, Bill Ward, and Al Cameron) found that Cameron's past results did display quite consistent relationships; for example, the fraction of the total colliding mass placed into orbit was maximized when the offset between the centers of the colliding objects was about 80% of that in a grazing pass.

Using these results, I was then able to make a prediction of what impactor size and impact angle would be optimal for producing the Earth–Moon system. Erik Asphaug and I simulated these specific types of impacts, and found that indeed they *do* appear to be able to simultaneously yield the Earth–Moon masses and angular momentum (see Figure 1 from Canup and Asphaug, 2001), while also placing sufficient iron-free material into orbit to yield the Moon. This class of impacts represents the least restrictive case, in the sense that no further modifications of the system mass or angular momentum are required after the Moon-forming impact. Ironically, the single impact that we "discovered" could produce the current Earth–Moon system was nearly exactly Mars sized: the impactor size that had originally been proposed by Cameron and Ward in 1976, a decade before the first impact simulations were performed. Early impact simulations results had implied that a larger impactor on a less-grazing trajectory was required to avoid placing too much iron into orbit, but when we re-examined the original "small" impactor cases with much higher numerical resolutions (now possible due to computer speed increases) we discovered that this was not the case.

Much work remains to be done. A new prescription for how molecular materials thermodynamically respond to release after impact has recently been developed by Jay Melosh (University of Arizona), and is currently being incorporated into the impact simulations. This will provide a greatly improved understanding of the degree of melting and vaporization in both the ejected material and the Earth. A careful reconciliation of the amount of heating implied by the impact simulations with the Moon's volatile element inventory has also yet to be completed. But in some respects, the giant impact simulations have now come full circle. The simplest scenario from a dynamical standpoint – but certainly not the only one imaginable – involves the impact by a Mars-size object with Earth 4.5 billion years ago, near the very end of Earth's formation. The fact that the single impact necessary to account for the mass of the Earth and the Earth–Moon system angular momentum also appears to be that needed to place the right amount of material into orbit to form the Moon provides strong support for the impact hypothesis – and for the notion that our Earth and Moon owe their paired existence to a chance, and perhaps rare, single event.

Moondance

The above story has covered only a portion of the history of one particular area of planetary science, but hopefully in doing so it has also provided a more general glimpse into the nature of scientific research. As an individual scientist, I do not know that I am particularly characteristic, except that many of us tend to be quite individualistic and creative sorts, prone to all sorts of varied activities. While Hollywood often portrays scientists as socially inept, reclusive types who are indifferent to humanist concerns, the reality is that most scientists are warm and intricate people, who live varied and well-rounded lives. Among the planetary scientists I know you would find pilots and painters, authors and musicians, history buffs and vintage-car enthusiasts, actors and activists.

In that vein, I will close on a personal note. Soon after I started graduate school, finding myself somewhat fatigued by the purely technical focus of my course work, I began dancing again. For the next ten years I danced and performed with Boulder Ballet, for most of those years as the company's principal dancer. During that time, I was able to dance many of the roles that I had dreamed of as a girl, and while balancing the demands of science and ballet was stressful at times (particularly while writing my dissertation!), I found it immensely enjoyable and rewarding.

I recently retired from my ballet career, but in science I am happy to find myself still a youngster, looking forward to many years of work and possible discoveries to come.

Further reading and references

Brush, Stephen G. (1996). *Fruitful Encounters: The Origin of the Solar System and of the Moon from Chamberlin to Apollo*, Cambridge University Press.

Canup, R. M. and E. Asphang (2001). *Nature*, **412**: 708–712.

Ida, R. S., R. M. Canup and G. R. Stewart (1997). *Nature*, **389**: 353–357.

Ward, Peter D. and Donald Brownlee (2000). *Rare Earth: Why Complex Life is Uncommon in the Universe*, Copernicus.

7

To rove on the red planet

Matthew P. Golombek, Jet Propulsion Laboratory

Matt Golombek is a geologist, raised and educated on the American east coast, but who settled at the Jet Propulsion Lab in sunny Southern California for his career. Matt was the chief scientist of the highly successful Mars Pathfinder lander/rover mission. In that role he not only guided the project science team and helped ensure that the mission's dramatic flight would be a success, but he also appeared regularly on television news programs including the Today show and Nightline, with an appearance on the Conan O'Brien Show. Matt's broad smile and ready wit make him a pleasure to be around, and a favorite of cruise ship entertainment directors around North America.

In the beginning of the project, I was not sure I even wanted the job; by the end, it was the peak experience of my entire career. At its start, Mars Pathfinder was defined as an entry, descent, and landing demonstration, with little or no science. What self-respecting scientist would possibly want to be the chief scientist on a mission without any science? By the end of the project, it was an amazing achievement by an absolutely committed, incredibly close, understaffed, and overworked team. It is an experience that could probably never be duplicated and one that I will never forget.

The Mars Pathfinder Project started in 1992 as a low-cost planetary mission that was to be done quickly and cheaply (a NASA Discovery mission). In a sense, it was a management test. Could JPL (Jet Propulsion Laboratory), which is NASA's lead center for planetary exploration, do a project "faster, cheaper, better?" It was conceived of as a demonstration flight of a network mission that would follow, in which 10–20 landers would be reproduced inexpensively and landed around the planet to measure the weather, seismicity, and sample the surface geology. Just after the Project actually got a new start, the loss of the Mars Observer spacecraft caused a complete reevaluation of future plans and the network mission was scrapped. However, even though Pathfinder was no longer connected to a subsequent mission, it was retained as one of NASA's first missions in this "new way of doing business." In addition, because it was in existence and somewhat separate from the Mars program that was subsequently defined, and because of its initial minimal science content, the Pathfinder project was largely ignored by both the science community and by JPL management.

Beginnings

I arrived at JPL about ten years before Pathfinder began, after a post-doctoral fellowship at the Lunar and Planetary Institute in Houston. Although my post-doc was a great job, it was temporary and I was happy to trade the swamps of Texas for the Mediterranean climate of southern California.

I was trained as a geologist, earning an undergraduate degree at Rutgers University in New Brunswick, New Jersey and a master's and Ph.D. at the University of Massachusetts, Amherst. I absolutely love doing research. I got my first taste as an undergraduate, where I did a project that showed that solid Earth tides triggered a volcanic eruption in Central America. My master's degree was on lunar grabens (fault bounded valleys) and my Ph.D. was a mapping and geophysical study of a normal fault zone in the Rio Grande rift in northern New Mexico. The task of wrapping your brain around a topic, of learning what is already known, integrating new observations, and discovering new aspects about the topic was what drew me into a research career.

I still remember a time as a graduate student working on the extended Viking Orbiter mission. It was December of 1977 and I went to the Jet Propulsion Laboratory (JPL) in Pasadena, California to work at the Viking project to obtain images of areas we were working on in the western hemisphere of Mars. By this time, the project was a pretty sleepy place run by a small group of engineers. The landings had occurred a year and a half before, and most of the scientists were back at their home institutions. I got to work with the Viking engineers directly on new sequences of images. I still remember watching the monitors with amazement a day or two after my requests and seeing the images of Mars, which was hundreds of millions of kilometers away, come down to Earth for the first time. Each new sequence of images uncovered areas of Mars in a detail never seen before.

Although I did not fully appreciate it then, JPL is an interesting mix of academia and business. JPL is predominantly run by engineers who build and operate spacecraft, which is JPL's prime business. Scientists are at JPL primarily to service missions. When I first got to JPL, it was a great place to do research. I got to collaborate with a wide variety of different scientists on a wide variety of topics. After being here for a number of years, I was asked to help out with flight projects that were under study. These early missions included the Mars Rover Sample Return, a Mars Network mission, and a variety of other rover, lander, and sample return missions that were being studied. These studies were mostly frustrating affairs that never got beyond the viewgraph stage, so lots of effort was put into paper and plastic film, but none got selected as a new start. By the time Pathfinder got its new start, it was a welcome change; Pathfinder was real, with real hardware, instruments, and a real launch vehicle that would take us to Mars in a very short time!

Academia and projects

One aspect of academia that dawned on me during my early years at JPL was how small the planetary research community is. There are probably a few hundred active planetary geologists in the World (most are in the United States). After working in the field for a dozen years, you get to know most everyone in your particular area of expertise and in the field in general. Research involves countless hours of working in relative obscurity on topics understood by a limited number of other scientists in that specialty. Some of us joke that we spend hours writing proposals to get funding to do research, carry out that research, and write papers reporting on the results that are actually read by a very limited number of other scientists. Furthermore, in planetary geology many of the topics that are worked on will not likely be tested for many years until another spacecraft gathers additional information. In fact the way in which planetary geologists get new information is typically by new spacecraft that are sent to the planet of interest. As a result, major advances in planetary geology generally require spacecraft to make new measurements that allow new research to be carried out.

Projects that build and operate spacecraft, however, are not academic enterprises and are nothing like academic research. Such projects require hundreds of people to build the spacecraft, they cost hundreds of millions of dollars (and typically much more) and take many years to complete (typically around ten years). Mars Pathfinder was a Discovery class mission, done "faster, cheaper, better." It took about 3.5 years from project start to launch, which is a factor of 3 less than typical for planetary missions. It cost between 200 and 300 million dollars to complete, which is comparable to the cost of major motion pictures and a factor of 10 or even 20 less than many other planetary missions. For an academic research scientist, being involved in a mission was a completely new experience. For the first time in my career, I was part of an enterprise much bigger than myself (or my research), that was important to hundreds of others, and for which I was a critical part, providing environmental information to the engineers and helping develop the scientific objectives and goals of the mission. This was a true team effort and I was a key member of the team. This was not some esoteric academic research project; this was a real engineering project that mattered to many hundreds of people.

A spacecraft

Robotic spacecraft are among the most complex devices built by humans, and launching and operating them in space are risky and difficult endeavors. Because of the disparate orbital rates of Mars and Earth about the Sun, spacecraft can be launched to Mars along minimum launch energy trajectories only once every 26 months. This is the ultimate deadline. If the spacecraft is not built, tested, and erected on the launch vehicle in time for the roughly one month launch window, it will be scrapped, and the project terminated, with no results. The spacecraft must be designed to traverse through space, get energy from the Sun, listen and respond to commands from the Earth, and to make observations with carefully designed and calibrated instruments, and to send those data back to Earth. These spacecraft require highly coordinated work by hundreds of different kinds of engineers. There are mechanical engineers to develop the structures and moving parts, electrical engineers to design the power and electrical system, as well as others to design the computer, the command and data-handling system, and the antennas and communication systems to name just a few. All of the components must be lightweight (launching something into space costs about $10,000 per pound), space qualified (designed to survive and work in space) and packaged to fit on top of a launch vehicle. All of these systems must work perfectly together, requiring extensive testing.

Space projects under development are managed by engineers, who make sure all of the components come together to meet the schedule and budget. Scientists work with the engineers to make certain that the spacecraft and its instruments can carry out the science objectives of the mission. Scientists are in a distinct minority during spacecraft development and there can be an adversarial relationship between engineers and scientists, because their concerns are different. Engineers build things that work together (like spacecraft); scientists analyze data to learn about an object or a process. Sometimes the scientists write down high-level science objectives that have been blessed by learned scientific groups and "throw them over the fence" to a group of engineers who do not fully understand what the scientists have written. As spacecraft are developed there are invariably trades that must be made that affect performance (ultimately the scientific measurements). In the world of spacecraft development, performance is constantly being measured against the real engineering constraints of the developing spacecraft. In the best of worlds, the scientists and engineers work together in a collaborative effort to do the best science under the constraints of the mission. In the worst of worlds, the engineers and scientists distrust each other and have difficulty working together.

An opportunity

Every project at JPL (and probably most places) develops a personality. The personality is set by the few people who start it, typically the Project Manager, the Project Scientist, and its top managers and engineers. It is often said that the personality is set early and survives even if some of the key people leave or move on. Pathfinder had great people who always tried to do the right thing. This project personality created a tremendous opportunity. So what if the project was defined as an entry, descent, and landing demonstration? It was still going to Mars, and no one wanted to carry red bricks. Why not make the best of the instruments and payload to optimize the science the mission would do? This attitude quickly spread and a real synergy developed between the scientists and engineers. The payload that Pathfinder was to carry was defined early. It had to take some pictures of the surface to show the public; it would carry a small rover to explore the nearby surroundings, perform a variety of technology experiments (that is, test rover technology), and measure the chemical composition of rocks and soil; and the lander would measure the temperature and pressure of the atmosphere during entry and descent and after landing. This top-level definition left plenty of room open about exactly how the instruments would be designed, built, and used, and what would be learned. This opportunity was what made the job so interesting. With the right attitude, work, and interaction with the engineers, the Pathfinder mission could be made into a legitimate science mission. No measurements of the chemistry of rocks on the surface of Mars had ever been made before and, if their mineralogy could be determined, the mission could be used as a "ground truth" for orbital remote sensing measurements.

Motivation

Mars is the most Earth-like planet in our Solar System. It is the first planet humans will visit and the only planet besides Earth with abundant water that can support life, including people, in the future. The geologic record on Mars suggests an early climate that was warmer and wetter in which liquid water (a requirement for life) may have been stable. Further, study of meteorites from Mars has suggested to some that they may contain evidence of primitive life. Mars is also a unique terrestrial planet with evidence for major climatic change and a geologic record of rocks on the surface that spans the entire history of the Solar System. As a result, an exploration program of the Red Planet will allow the investigation of a wide variety of fundamental geologic, climatologic, and exobiologic questions. Was Mars once warm and wet like the Earth? Did life begin on Mars (if not, why not)? If it did, what happened to it? Could it have come to the Earth in chunks of rocks ejected from Mars during violent cratering events? Could we all be Martians? Is life common or rare in the universe? And, a question of almost theological importance: are we alone in the universe?

With the right landing site, the rover and imager could potentially learn something about the early environment on Mars. If rocks could be identified in terms of their type and mineralogy, then the environment in which the rocks formed could be determined. Effectively, the Pathfinder mission was a rock mission and the rover was a one-foot tall field geologist. A well-trained field geologist looking at and identifying rocks can determine the geologic history. The Pathfinder rover would study rocks up close and try to learn about the environment that existed when those rocks formed. All that was needed was the right landing site.

A landing site

Where should we land on Mars? This is one of those questions that many scientists and engineers think they know the answer to. "Sure, let's land here and look at this or that." Many considered themselves experts, even with little thought about what is involved in selecting a landing site. Nevertheless, selecting the right landing site was a critical decision for the Pathfinder project. The entire success of the mission rested on selecting a place that was safe for the lander. Obviously, if the spacecraft does not land safely there is nothing to show for the effort (including no science). The location must be one where the payload can address fundamental scientific questions, so that the landing site must be both safe and scientifically interesting. Everything that can possibly be learned about the site must be, as the fate of a spacecraft that cost a couple of hundred million dollars rests on this decision. It is one thing to write a science paper about some topic, it is something else entirely to risk an entire mission on the interpretation. This was not an exercise for the faint of heart. It was decision so important that as project scientist I did not feel I could delegate it to anyone else.

So how does one pick a landing site? First the lander and how it comes to rest on the surface must be thoroughly understood. Close collaboration with the engineers is required to understand what characteristics of the surface are important to the landing system. For most landers, it is elevation, latitude, and the physical properties of the surface. For landers on Mars that use the atmosphere to slow the spacecraft via an aeroshell and parachute, a certain atmospheric column density is needed to bring the vehicle to the correct terminal velocity. This translates into an elevation for the season and time of arrival, with lower elevations generally being better. For solar powered spacecraft, sites near the subsolar latitude (within 25 degrees of the equator) are preferred for maximizing power that is used to withstand the strong daily thermal cycling at the surface. In addition, the surface needs to be competent for landing and roving, without under-dense material, slopes which are too steep, or with too many rocks. Finally, from a scientific point of view, the geologic setting determines what scientific questions can be addressed.

The biggest problem was that the lander and rover were susceptible to meter scale hazards (rocks and slopes). Yet the highest resolution images that existed were 20-year old Viking data that could only resolve features about a kilometer in size, and there was no way to get any other images of higher resolution. After a two-year process that included two open landing site workshops and multiple reviews, the project selected a landing site downstream from the mouth of a giant catastrophic outflow channel, called Ares Vallis. This site offered the potential of identifying and analyzing a wide variety of crustal materials, from the ancient heavily cratered terrain, intermediate-aged ridged plains, and reworked channel deposits, and provided a calibration point or "ground" for orbital remote sensing observations. Such information would address first-order scientific questions, such as differentiation of the crust, the development of weathering products, and the nature of the early Martian environment and its subsequent evolution. The site was believed to be safe based on extensive analysis of existing remote sensing data and by analogy with the Channeled Scabland in Washington state, believed to have been produced by a similar process of catastrophic flooding at the end of the Ice Age. Our second landing

Figure 7.1. Drawing presented to the author by teachers at the end of the *Mars Pathfinder Landing Site Workshop II: Characteristics of the Ares Vallis Region and Field Trips to the Channeled Scabland*, September 24–29, 1995 in Moses Lake and Spokane, Washington. The teachers obviously recognized the difficulty of predicting the meter-scale characteristics of a surface on Mars from kilometer-scale remotely sensed data. This workshop is described in two technical reports M. Golombek, K. Edgett, and J. Rice, eds., *Mars Pathfinder Landing Site Workshop II: Characteristics of the Ares Vallis Region and Field Trips in the Channeled Scabland, Washington*, Lunar and Planetary Institute Technical Report 95-01, 1995, Part 1, 63 pp; Part 2, 47 pp., and in a news item, K. Edgett, J. Rice and M. Golombek, Scientists, Educators Prepare for *Mars Pathfinder Mission*, EOS, Transactions, American Geophysical Union, v. 77, pp. 9–10, January 9, 1996.

site workshop, in Spokane, included field trips into the Channeled Scabland and involved an equal mix of scientists, engineers, and science teachers from Washington and Idaho. By the end of the workshop and field trips, the teachers grasped the difficulty and importance of predicting the nature of the Martian surface from the data available and produced the drawing that still hangs on my office wall (Figure 7.1).

Risk

Any act of exploration involves some element of risk. Risk comes from exploring the unknown. Any space mission involves risk of both unknown worlds and environments and also the risk of traversing through space. Space is an unforgiving environment and even getting into space involves a real chance of complete destruction during launch. As an example, Mars Pathfinder was launched on the world's most reliable expendable launch vehicle. In one hundred launches, it explodes a handful of times. This fact hit me directly in the cold early morning darkness of Cape Canaveral Air Station in Florida as I watched the launch of Mars Pathfinder, which I had spent five years of my life building. Two weeks later, the same launch vehicle exploded, destroying the satellite it was carrying.

Mars Pathfinder involved more risk than most space missions. Clearly one significant risk involved just getting as far as the launch pad. The project was an impossible task. We had just over three years from project start to design, build, test, and launch a spacecraft, all for about 150 million dollars. For reference, the Viking missions to Mars cost about 20 times more and took approximately three times longer. In reality, no one actually knew if the job could even be done. The project was under incredible fiscal and time constraints from the start and we were hounded by reviews throughout the entire process. Spend too much or take too long and get cancelled. These constraints required new ways of doing business, which included assembling a small, hand-picked team from JPL, and co-locating them away from the technical divisions. If someone was not up to the job they were let go. To those of us within the project, it felt like it was "us against the world." This created a team atmosphere and an incredibly committed workforce. We were understaffed, which offered opportunity – if someone saw a problem, they fixed it, even if it was outside their own area.

Having too few people actually was an advantage compared to having too many, which can result in over compartmentalization and poor communication. The project was small enough so that we actually knew most everyone who worked on it. For the scientists working on the mission, we simply rolled up our sleeves and worked side by side with the engineers to solve problems. We were all empowered to make decisions and we were all committed to do the impossible.

Of course, getting to the launch pad was only part of the problem. The spacecraft had to get to Mars and land safely, and landing is much more difficult than simply going into orbit. Our scheme was to enter the atmosphere directly, traveling at about 27,400 kilometers per hour, and use the atmosphere to slow the lander down before inflating giant airbags to cushion the final impact. This would take about four minutes and in that time about 70 pyrotechnic devices must trigger a series of events that must work perfectly for the lander to a have any chance of surviving. Naturally despite careful design and significant testing, prior to landing there was plenty of skepticism regarding the mission's chance of success.

As if these risks were not enough, there was career risk as well. Would you spend five years of your life consumed by a spacecraft that most people thought would have little chance of making it to the launch pad and even less chance of landing successfully on Mars? The career risk for a scientist involves losing touch with their research. Project scientists are selected because they are respected researchers in their fields. But the job of being a project scientist does not leave much time for research. My research papers during the project were partly limited to those that were almost finished before the project started. This is clearly a problem for scientists on soft money who must maintain a vigorous publication record to be successful at obtaining grants to pay their salary (publish or perish). To make matters worse, I was risking my research career on a mission that many planetary scientists thought (incorrectly) would have little science payoff.

Of course, the success of Pathfinder paid off in many ways (Figure 7.2). Among them were two bets I made during Pathfinder. One concerned whether we would be able to get the spacecraft to the launch pad with the payload complement as originally defined. I won that bet; it was a bottle of fine California wine that I received in public at the Planetary Geologic Mappers meeting held after landing in July 1997. The second

Figure 7.2

Plaque given to the author after his second presentation to the Coast Geological Society, a professional organization of geologists in the Ventura area of California.

bet concerned the nature of the landing site, which received some press (Kerr, 1996). On the basis of the remote sensing data and the Earth analog (Channeled Scabland), I predicted the site would be a depositional plain composed of materials deposited by the catastrophic flood (others believed it had been subsequently covered by volcanic rocks). This bet was for a glass of beer that I collected at the March 1998 Lunar and Planetary Science Conference. It was right up there with the best beer I ever tasted.

Develop and test, test, test!

Too little time and too little money forced the Mars Pathfinder project to use pre-existing space qualified hardware wherever possible (which tended to drive the spacecraft mass up), to contract out subsystems to industry, where its expertise exceeded ours, and to thoroughly test the spacecraft and subsystems. Every subsystem was extensively tested. I often joked that the engineers really had fun throwing things out of aircraft during these tests. The short development time required design activities concurrent with testing. As a result, if a test showed a problem, solutions would ripple back through the spacecraft design and fabrication activities. This created a very dynamic environment. Towards the end of spacecraft development I distinctly remember project personnel's psyches being crushed or buoyed by test failures or successes.

Two key tests on the airbags and solid rockets came right down to the wire. The airbags were tested in the largest vacuum chamber in the world in Sandusky, Ohio. A full-scale lander with inflated airbags was placed inside the chamber, which was pumped down to Martian atmospheric pressure. A large bungee chord accelerated the lander to over 25 m/s impacting against a 60 degrees dipping platform covered with extreme concentrations of angular, sharp volcanic rocks up to 0.5 meters high. Initial airbag drop tests showed that the airbags tended to tear against the rocks, catastrophically losing pressure. The engineers kept adding abrasion resistant outer layers sewn over an interior bladder to try and maintain airbag pressure. Each failure would require new tests. Near the end, the project had enough time (and budget) for one more series of drop tests, in which three layers of abrasion resistant material were sewn over the interior bladder and tested at the maximum anticipated landing velocity (that the previous airbag test had failed).

The backshell solid rockets were designed to remove any residual vertical velocity just before landing. The decision was made early on to use low aluminum propellant rockets to minimize potential contamination of the landing site (the initial design had vented airbags indicating the lander might come to rest near where the solid rockets fired). Final solid rocket testing when fired attached to a platform showed dynamic instabilities and the loss of a nozzle. There was only a few days time to make the decision to use the higher aluminum rocket propellant and try one last test.

We almost lost the spacecraft!

The spacecraft was launched on December 4, 1996 at 1:58 a.m. eastern standard time from Cape Canaveral Air Station in Florida. It was the third day of the one month launch period, in which the first attempt was canceled (without a countdown) due to bad weather and the second attempt was scrubbed a few minutes before launch due to a problem with non-synchronous computers at the launch site (the launch window was one minute each day). About one and a half hours after launch, Pathfinder exited the Earth's shadow, with the Sun illuminating the cruise-stage solar panels and Sun sensors for the first time. Current began flowing from the gallium arsenide solar panels on schedule, but two redundant Sun sensors did not detect the Sun.

These sensors each provided two axes of Sun angle data to the onboard attitude control software and were needed to carry out turns and course corrections during cruise, which were critical for getting to Mars. Within days it was determined that both Sun sensors had been optically blocked, one partially and the other fully, by residue from a pyrotechnic detonation that had occurred on the upper stage of separation during launch. Spacecraft engineers worked around the clock during a critical (and tense) period to modify the software to recognize the reduced-quality information that was available from the partially obscured Sun sensor. These changes allowed operation of the attitude control system and the necessary turns and course corrections during cruise.

Surface operations testing

After launch we had to train and organize ourselves to operate the spacecraft efficiently. A series of Operations Readiness Tests (or ORTs) were designed (Figure 7.3). These tests ranged from individual software tests of the flight computer during specific entry, descent, and landing events to full dress rehearsals of landing and surface operations with the entire operations team (including scientists) on Mars time. A Mars day, or Sol, is 37 minutes longer than a day on Earth, so that operations activities synched to daylight hours on Mars (required for a solar powered lander and rover) advances or "walks" forward 37 minutes each day. Some of the ORTs were dramatic failures. During one entry, descent and landing software test, the computer reset so the heatshield and backshell did not eject and the parachute did not open, although the solid rockets did fire. On one of the full up dress rehearsals, the rover could not be driven off the lander petal for a number of Sols, thereby depriving the lander of solar power, depleting the battery, and ending the mission.

Organizing the science team and leading them in an integrated fashion required to operate the rover was also a challenge. Organizing scientists has often been likened to herding cats. We had close to 80 scientists working on the mission at landing. They each had their own opinions and ways of working. We had to go through detailed strawman landing operations scenarios involving what data would be acquired when and what data would be downlinked when to support rover and surface operations. My approach was to continually impress upon them that everyone benefited from carrying out the best mission that could be done. This approach of providing a focus and vision appeared to eliminate the political in-fighting that can occur in a diverse and competitive science team. This philosophy also went towards releasing virtually all data as soon as it was received, an approach that helped fuel the mission's popularity. During the nominal mission, three shifts worked around the clock; each team worked during a particular period of the Martian day, which shifted forward about 24 hours during the first month. The pace of activity and "living on Mars time" exhausted the operations team by the end of the

Figure 7.3

One of the first mosaics of the rover stowed on the lander petal by the lander imager in a test facility at JPL produced by the author early in the Mars Pathfinder operations testing program at JPL. The mosaic was put together crudely (note image borders) by placing individual lander images on the computer screen at roughly the correct position. Much more sophisticated mosaics and tests were subsequently carried out by the operations team to be prepared for carrying out surface operations.

nominal mission (in August 1997), and we shifted to a less intense phase of operations commensurate with the extended mission.

Success

None of us will forget the somber mood in the mission control room in the Space Flight Operations Facility (SFOF) at JPL that morning; we had all worked on the project for about five years and knew this was it. The previous night at 8 p.m. and again at 3 a.m. Pacific Daylight Time on July 4 we had reviewed the tracking data from the spacecraft and evaluated whether or not to do an emergency trajectory correction maneuver. The navigation data indicated that we were approaching the southwest portion of the landing ellipse, where a large and potentially hazardous hill was located. We did not make the emergency trajectory correction maneuver, because the uncertainty associated with the maneuver was potentially greater than the possible hazards we might encounter. The landing sequence began at about 10 a.m. and we all listened intently for the reports of the signal from the spacecraft. We knew that the cruise stage had separated and that Pathfinder had entered the atmosphere, but communications were interrupted (as expected) during initial entry. Most of us had not really expected to hear again from the spacecraft during descent and immediately after landing as the first transmission of data was not scheduled for several hours, until the lander had deflated and retracted its airbags, and opened its petals. Exhaustion was replaced by exhilaration as we received tones sent by the spacecraft during descent on the parachute and after the lander had come to rest, still within its inflated airbags. It had worked perfectly. We had done it and we cheered our success.

Mars Pathfinder landed safely at Ares Valles on July 4, 1997, thereby demonstrating a robust and inexpensive landing system and a small rover that collected data from three science instruments and ten technology experiments (Figure 7.4). The mission operated on the surface of Mars for three months and returned 2.3 Gbits of new data, including over 16,500 lander and 550 rover images, 16 chemical analyses of rocks and soil, and 8.5 million individual temperature, pressure and wind measurements. The rover traversed 100 meters clockwise around the lander, exploring about 200 m^2 of the surface. The mission was also a scientific success (Figure 7.5, see color plate section). It returned high-quality scientific data that addressed important science topics such as the geology and geomorphology of the surface, mineralogy and geochemistry of rocks and soils, physical properties of surface materials, magnetic properties of airborne dust, atmospheric science including aerosols, and rotational and orbital dynamics of Mars. The science results have been described extensively elsewhere, but data returned by Pathfinder have significantly changed our understand-

Figure 7.4

Mosaic of the rover stowed on the lander petal after landing safely on the surface of Mars at Ares Vallis on July 4, 1997. Compare with Figure 7.3.

ing of Mars. Taken together, the rounded pebbles, cobbles, and the possible conglomerate, the abundant sand- and dust-sized particles and models for their origin, and the high silica rocks, all appear consistent with a water-rich planet that may be more Earth like than previously appreciated, with a warmer and wetter past in which liquid water was stable and the atmosphere was thicker.

The Pathfinder landing grabbed the attention of the entire world. The mission captured the imagination of the public, garnered front page headlines during the first week of mission operations, and became one of NASA's most popular missions. A total of about 566 million Internet "hits" were registered during the first month of the mission, with 47 million "hits" on July 8 alone, making the Pathfinder landing by far the largest Internet event in history at the time. We became celebrities overnight, appearing on the nightly news with unprecedented media coverage. We were swept up in a maelstrom of publicity and public reaction that left little time for sleep during the first month of surface operations.

No one could have predicted the public's reaction to Pathfinder. What made it so popular? What led to its success with the public? First of all, landing on a planet is a defining moment that is tailored for media coverage. It is a "do or die" moment that most everyone can understand and relate to. July 4, a national holiday, is typically a slow media day, so there was little news competition to our landing on Mars. Further, an entire generation had never witnessed a landing on another planet. It had been almost 20 years since the Viking landings and they were overshadowed by the human landings on the Moon that had just occurred. Third, having a rover that went out and explored kept people's attention for days beyond what a stationary lander would have. Fourth, we released virtually all data that we received immediately on a system of web sites that allowed people to explore Mars on their own in much more detail than they could via radio, newspaper, or television reports. Finally, the country could see that the team was composed of real people who were absolutely committed to the mission. There was no political imperative behind Pathfinder and it was not a military style conquest. It was the equivalent of a group of people constructing a spacecraft in their garage for no other reason than because they loved building it, exploring Mars, and learning about our neighboring world. In a way, Pathfinder became a moment when humanity came together and marveled at the exploration of a new and different world, for no other reason than it was amazing and a tribute to what a space program can accomplish in the best of situations. To be a key participant in such an event is rare and one I will always treasure.

Further reading

Golombek, M. P. (1998). The Mars Pathfinder Mission, *Scientific American,* 279: 40–49

Kerr, R. A. (1996). Gambling on a Martian Landing Site, *Science,* **272**, April 19: 347–348.

Raeburn, P. and M. Golombek (1998). *Mars: Uncovering the Secrets of the Red Planet,* Washington, DC: National Geographic Press, 231 pp.

8

To the asteroids, and beyond!

Richard P. Binzel, Massachusetts Institute of Technology

Ahhh, Rick Binzel. Rick and I have known each other since he was an entering grad student at the University of Texas and I was a finishing undergraduate. Rick is among the world's premier asteroid observers, the man who led much of the mapping of Pluto during its eclipses with its satellite Charon, the originator of the Torino asteroid impact hazard scale, and one heck of a nice guy. He's also a connoisseur of fine wines and ports. He and his wife Michelle have two children, whom they are raising near Boston and (more recently) in Paris.

"Can you see Jesus?" came the question out of the darkness. Although the voice was soft and gentle, it startled me nonetheless. Tucked in the shadows of a small grassy area behind my parents' garage, I was concentrating so hard on the star-filled view through my small telescope that I had not heard the approaching footsteps. "Well, not exactly" was the best response I could come up with after managing to recapture my breath. Peering through the darkness I saw the outline of a long-haired young man decked out in crosses and beads, a vintage 1973 "Jesus freak." I recognized him as a member of a group renting a small house down the alley running past the garage. I explained what I was doing, as simply as I could. Gazing at the night sky we marveled together over the magnificence of Nature. While it was a pleasure to share that indescribable sense of awe with another human being, I was delighted when he turned toward home, leaving me to the solitude of my task. You see, I was taking my seventh grade science project seriously. I was charting the path of the asteroid Vesta through the late winter sky.

Space age kid

My seventh grade asteroid tracking project was just one part of my childhood interest in space. In fact, I can not remember a time when I did not want to study space. The 1960s and early 1970s were an inspiring time for an aspiring young scientist. I could name every piece of every rocket used in the US manned space program. I begged (or feigned illness) to stay home from school to watch every minute of the Apollo moon walks. My parents weren't fooled (my father is a physician!), but they let me watch nonetheless. I was not particularly in awe of the astronauts themselves – other than to admire their prerequisite coolness under pressure. It was the "guest scientists" sitting along side the network anchors who captivated me the most. They knew everything about outer space and the scientific experiments the astronauts were conducting. Dr. Harold Masursky is the one I remember best and I distinctly planted the idea in my ten-year-old brain: "Wouldn't it be the coolest thing to know everything about outer space? I want to be like that when I grow up."

I began reading through every astronomy book that our small town (Washington Court House, Ohio) library had to offer, and there were quite a few. My favorite reading spot was up in a tree house that I had built out of two-by-fours salvaged from the local lumber yard, with the plywood provided by Santa Claus. (Yes, my Christmas list contained both star charts and plywood!) My grandfather, a retired engineer who had graduated from MIT, was especially inspirational in sending me books, and posing and answering science questions. I began to learn about the big bang model of the Universe, how the nuclear furnace inside a star works, how stars clump into clusters and galaxies, and how shining clouds of hydrogen gas are the nurseries where new stars are made. The distances astonished me (and they still do), and perhaps the immensity was a bit too overwhelming for a small-town boy. Somehow, I always felt the most comfortable and the most intrigued when reading about the neighborhood around my tree house: our Solar System. Here, my grandfather told me, were worlds that were within reach of mankind's exploration during *my* lifetime.

Hooked on planets

One night it seemed the Solar System reached back. It was early December 1970 when my sixth grade teacher, Mr. Rodger Mickle, asked

Figure 8.1

The author at age 13 with an amateur telescope.

for a volunteer to accompany him on an evening shopping trip for the classroom Christmas tree. My hand shot up and I got the nod. I had received my first telescope for my twelfth birthday a few weeks earlier and I wanted to show it off when he came by that evening. (Mr. Mickle was the first science teacher who let me write reports on whatever *I* wanted, placing him in my Pantheon of great teachers.) Alas, the Moon was not up that evening leaving me wondering where to point the telescope? I dug out the Sunday *Columbus (Ohio) Dispatch* newspaper and found the weekly "sky watchers" column reporting the planet Saturn could be viewed as a yellow star in the east. Racing outside into the cold air I struggled with my new telescope to get it pointed at the brightest and yellowest star I could see rising over our neighbor's garage. Bringing the tiny image of Saturn and its rings into focus, I beheld an intensely real and captivating view that was stunning beyond words! It all became clear. Not only did I want to be a scientist who studied space, I wanted to be an astronomer who studied planets.

For my parents, it was the beginning of the end. Being no longer content just to read books, I began attending regular meetings and star

Figure 8.2

The author at age 40 at the controls of the Palomar 200-inch telescope. The asteroid Vesta appears on the TV monitor.

parties held by the Columbus Astronomical Society. "Attending" meant one or both of my parents would drive me 40 miles north to Columbus and either remain there or come back late at night or early in the morning to retrieve me. How they accommodated these trips within our busy family life (I am the fifth out of six children) is a mystery, the answer almost certainly involves aspects of special relativity theory that Einstein never dreamed of. With seeming cheerfulness, my parents also gave up the back room of our garage for my own "observatory," and acted pleased when I painted the room kumquat orange, a color I chose because the paint store was willing to part with this mis-mixed gallon for only a dollar. From this room, it was only a few steps out of a side door to the small grassy area that became my regular observing spot. Naturally such a fine observatory required increasingly better telescopes and I worked my way up to a new six-inch (15 centimeter) reflector, paying for it using money I earned from my summer lawn mowing and winter snow shoveling jobs.

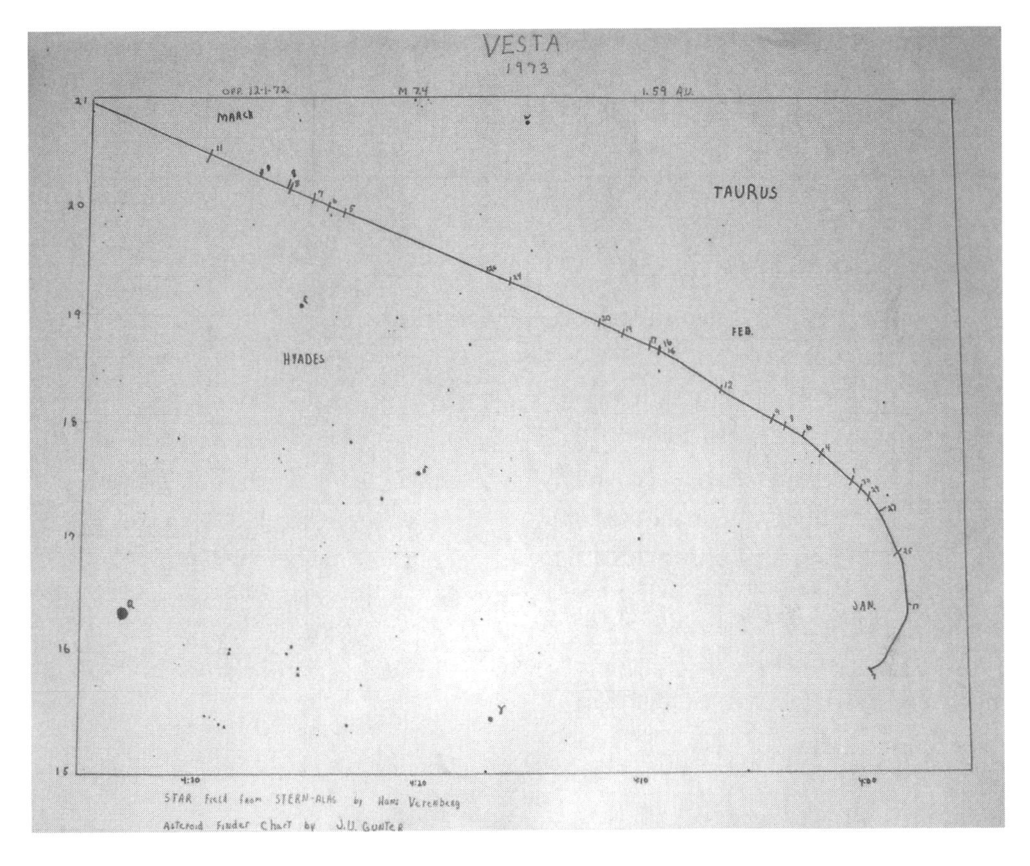

Figure 8.3 Seventh grade science project poster depicting the path in the sky of the asteroid Vesta.

Just as Saturn had taken me by surprise, so did asteroids. Working my way through the traditional rite of passage for amateur astronomers (spotting all one-hundred or so star clusters, galaxies, and nebulae on the list compiled by Charles Messier in the eighteenth century) I became eager to try other challenges to my observing skills. I read about a small newsletter called "Tonight's Asteroids" written by a retired pathologist living in North Carolina, J. U. Gunter. Amazingly, I learned that it was possible to spot these tiny worlds using a small telescope. Sending in my subscription of a few self-addressed stamped envelopes, my first newsletter arrived with a cheery note from Dr. Gunter penciled in the margin pointing out Vesta as being particularly bright and well placed for my first target. The hunt was on! This was not a distant galaxy residing in the same spot in the sky since the time it was first seen a few centuries ago: this was the second largest of all asteroids making an appearance in the sky as a slowly moving tiny star! Spotting it required good skills and a good star chart to examine the right star field at the right time. On January 25, 1973, in a star field not far from the star Aldebaran (the "eye of the bull" in Taurus), there was an "extra" star in the field just as Gunter's chart had predicted. It was Vesta! What's more, I could see it moving slightly northeast over the course of an hour! There was something captivating about watching this silent distant world move through space, leaving me wondering: What is this world like? Over the next two months I tracked the location of Vesta, overcoming the obstacles of bad weather, homework, and the occasional returns of a friendly Jesus freak. My poster board detailing my observations of Vesta earned me an A for my seventh grade science project. My career in asteroids was off to a solid start!

The making of a scientist

While tracking asteroids was a fascinating sport (all told I bagged several hundred different ones over the course of a few years), I wanted to study them in a scientific way that was beyond verifying the accuracy of their orbits by checking their positions. During an astronomy short course at the Columbus Astronomical Society, I learned about an instrument called a photometer – a device used to measure accurately the brightness of a star being viewed through a telescope. The application of this device was particularly fascinating for measuring asteroids because they change in their brightness over the course of minutes and hours. Many asteroids are elongated objects (picture an elongated river rock or a cigar) that are spinning consistently (rather than tumbling) about their middle points. As the asteroid spins, sometimes the broad side is reflecting sunlight back to Earth and sometimes it is just the narrow end. Measurements of the brightness of the asteroid made repeatedly over the course of a few hours reveals these variations and a diagram of these variations over time is the object's "lightcurve." Because the asteroid has four sides (two broad sides and two pointy ends), one complete rotation of the asteroid produces a lightcurve with two peaks and two valleys. Amazingly, even though the asteroid appears as nothing more than a tiny point of light when viewed through the telescope, using the photometer to measure the lightcurve can tell you how fast the object is spinning. One rotation period (that is, one complete day–night cycle) on most asteroids is only six to ten hours long.

Building a photometer required a high voltage power supply and a light-sensitive but expensive type of vacuum tube called a photomultiplier – all of which were beyond the reach of an eager 14-year-old in 1973. However the opportunity to use a photometer came in 1974 while I was attending a two-week astronomy camp for high school kids in California, Camp Uraniborg. The camp's director, Joseph Patterson (now an astronomy professor at Columbia), can probably count more professional astronomers today as his "students" than any other teacher in the world. (At least four of the campers from my year are now Ph.D. astronomers, including Neil DeGrasse Tyson, the director of New York's Hayden Planetarium.) Pairing up with Doug Welch (now a professor at McMaster University), we used the camp's photometer attached to a 14-inch telescope to measure the brightness variation of the asteroid 18 Melpomene as it rotated. Night after night we made repeated measurements of this small world, about 150 kilometers across and 200 million kilometers away, and found the lightcurve repeating every 11.5 hours. More than a century after this world was discovered, we were the first to deduce its rotation period. Writing together, the author team of Welch, Patterson, and Binzel published the results, giving me my first scientific paper at the age of 15.

There was no stopping me now, on my way to a professional career in astronomy. Bulldozing my way through high school in three years and graduating in 1976, I was attracted to Macalester College in St. Paul, MN. The college had an incredibly well-outfitted observatory, thanks to the wizard-like telescope-making ability of the college's astronomy professor, Sherman Schultz. My goal was to build a photometer system to carry on new asteroid observations, and, as if by magic, just the right types of telescopes (a 12-inch and a 16-inch both with Cassegrain optics) were constructed. Working together with another Macalester student (Patrick Hartigan, now an astronomy professor at Rice University), we measured lightcurves of both asteroids and variable stars, publishing our findings as we went along.

When it came time to head to graduate school for my Ph.D. training in 1980, the attraction of an astronomy department with lots of telescopes led me to the University of Texas in the scenic lake and hill country town of Austin. The University's McDonald Observatory, located in a remote mountainous part of west Texas near the town of Ft. Davis, provided an outstanding facility and dark skies for astronomical research. Professor Harlan J. Smith, the Observatory's director, provided the guidance and scientific training I needed to become a professional. Professor Smith gave me the discipline to define the exact scientific questions I was trying to answer within my Ph.D. research project. As a part of this project, I was measuring the range of rotation rates of asteroids, grouped together in clumps called "families." Was it correct that each of these families is comprised of the leftover pieces of a once larger asteroid, destroyed by a catastrophic collision? What clues may be left behind to reveal that these groupings shared a common origin? Using the skills I had developed as an asteroid observer, and many new ones to be sure, I worked for more than 300 nights to measure the rotations of more than one hundred asteroids both inside and outside several family clusters. Sure enough, my findings showed that the largest members within a family tend to be spinning at similar rates with their rotation poles pointing in similar directions. If a family was just a random collection of asteroids, there is little chance these properties would match at all. Indeed, these properties were random outside the families. Here, revealed by my photometer measurements, was tell-tale evidence of a common origin for asteroid families in violent collision events that happened many millions, if not a few billion years ago. Seemingly transcending time, through these present-day observations I was able to reveal key insights to the past.

Close encounters of the ninth kind

Something funny happened on the way to my Ph.D. that expanded my interest to the edge of the Solar System. Part of my graduate research training involved working for a McDonald Observatory research scientist,

J. Derral Mulholland, who had funding for a program to monitor the rotation lightcurve of Pluto, which repeats itself every 6.4 days. (A day–night cycle on Pluto lasts 6.4 Earth days.) Calculations in the early 1980s had shown that the orbit of Pluto's moon, Charon, would become "edge-on" as seen by Earth-based observers. Such an alignment only occurs for a relatively short period of time every 124 years! By using a photometer to measure how the combined brightness of Pluto plus Charon changed as Charon passed in front (a transit) or Charon passed behind (an occultation), we could actually measure new details about the surfaces and sizes of these distant worlds.

Looking at Pluto–Charon was like *déjà vu* all over again. Just a few years earlier, in the summer of 1978, I had an internship at the US Naval Observatory in Washington, DC. Returning after lunch on June 22, there was a small group of staff members huddled around a table in Robert Harrington's office next door. Either Bob or his good friend James Christy saw me walking by and called out, "Hey kid, come have a look at this." Sitting on the table was a binocular microscope used to examine photographic images taken on glass plates. Peering into the eyepiece I could see an elongated image of Pluto. It was just the sort of distorted blurry picture you would get if someone jostled the telescope or the camera while it was being taken. Being appropriately unimpressed, I was then instructed to examine the entire plate. Therein were a total of six separate images of Pluto with the same bumpy shape as well as images of background stars whose perfectly round shapes indicated that nothing had gone bump in the night during the exposures. What was the cause of the bump on Pluto? Could it be a moon, I asked? The discoverer, Christy, did indeed believe that this was evidence for a large satellite of Pluto. When confirming observations were obtained just a few weeks later, the discovery of Charon was announced to the world.

Working at the University of Texas in 1982, one of the most amazing things to me was that this once per century opportunity (that last occurred near the time of the American Civil War) was happening almost immediately after Charon was discovered. If Jim Christy had not made his discovery, would these unique events have gone unnoticed? Faced with great uncertainty in their sizes and in the exact orientation of Charon's orbit around Pluto, we were highly uncertain as to when we would begin to see these "eclipses." (Our generic but technically incorrect term for transits and occultations.) There was even skepticism by some astronomers as to whether Pluto's moon really did exist. Thus each trip to the telescope (a 500 mile westward journey from Austin) to monitor Pluto was one of great excitement and possible importance. Three years into the search (which, yes, was beginning to feel like a grind) brought the frosty, crisp, and crystal clear morning of February 17, 1985. Using the McDonald 36-inch telescope, I was working on asteroid photometry for my Ph.D. thesis until 4:00 in the morning when Pluto rose high enough in the southeast for me to turn my attention to it.

Being impatient, ahead of time I pointed the telescope as close to the horizon as it would go and waited for Pluto to come into view. While Pluto was rising above the bottom slit of the telescope dome, I used my finder chart to check the star field to identify which "extra star" was Pluto in the same way I had first spotted asteroids from my backyard just ten years before. I found Pluto easily and at last it had risen high enough to provide the telescope with a clear view through the dome slit. Giving the instrument one final check, I flipped the switch on the data acquisition computer (with a whopping 4K of memory) and began measuring the brightness of Pluto.

From our previous years of monitoring work, I knew that Pluto was normally quite constant in brightness during this part of its 6.4 day lightcurve. This night, however, something different was happening. Instead of being constant, Pluto's brightness started decreasing by just a few percent. Such a change was easily measured by the photometer and easy to calculate in my head (and confirmed by punching the numbers into my TI-59 calculator). After an hour or so, Pluto began returning to its normal brightness. The eclipses between Pluto and Charon had begun! The existence of Pluto's moon was confirmed! Checking and rechecking my numbers after sunrise, I called Harlan Smith to tell him the news. He asked, "Are you sure?" My answer of "yes" was proof enough for him. Just a few hours after I finally went to bed for my daylong sleep came a knock on the door. One of the cooks said I had a telephone call. Complaining all the way up the stairs to the kitchen I answered the phone and found it was a reporter from the *Washington Post* who snappily responded to my complaint with my first lesson in media relations: "Sleep doesn't matter," she said, "reporters have deadlines!" My observations of Pluto and Charon were front page news! The ensuing media attention was both thrilling and exhausting and I was relieved when the attention subsided a few weeks later and I could continue my work without interruption.

The principal work at hand was to use the transits of Charon in front of Pluto to map features on its surface. This mapping work, done by Eliot Young under my supervision, was achieved by comparing whether the dimming in the lightcurve was proportional to the area covered. In some cases we found that the dimming was unusually strong, even though the covered area was small. Our interpretation for this unusual cause and effect was that the covered location on Pluto must be particularly bright. In other cases we found a large covered area on Pluto resulted in only a small reduction in its measured light. For this circumstance, the covered area must be relatively dark. (Blocking it from view had little effect on the total brightness, since being dark, it did not contribute much to the total.) Interestingly, Eliot and I found that the bright regions on Pluto were at the poles and the dark region occupied the equator. We discovered that Pluto has polar caps! It seems most likely that these bright polar caps can only be explained if Pluto has "seasons." Most likely, these seasons are related to Pluto's highly elongated orbit around the Sun. The drop in temperature that Pluto experiences as it recedes from inside Neptune's orbit to its furthest point from the Sun (a distance ranging from 30 times to 50 times the Earth–Sun separation) likely leads to a significant amount (perhaps all?) of its thin atmosphere becoming frozen on the surface. Our best guess is that a build-up of this frozen atmosphere at the poles creates the bright polar caps. Pluto's tilt (it lies nearly sideways as it orbits around the Sun) probably contributes to making this buildup of ices (probably composed of nitrogen and methane) occur at the poles rather than elsewhere on its surface.

A calling to Vesta

Just as Pluto had seemed to come back and catch me by surprise, so too did Vesta, the second largest asteroid. Arriving at MIT as a young Assistant

Professor in 1988, I had a new telescope to use: the 2.4-meter Hiltner telescope located at the Michigan–Dartmouth–MIT (MDM) Observatory on Kitt Peak in Arizona. There the workhorse instrument was a spectrograph, a device which takes the light gathered by the telescope and spreads it out into its component colors. The colors of asteroids (and hence their spectra) differ from one another, depending upon their composition. Some are reddish and rocky, being composed of minerals with names like olivine and pyroxene that are common in rocks on Earth. Others are deep black and composed largely of carbon. Shiny gray asteroids, not surprisingly, are largely composed of metal. The scientifically interesting questions include: Why do asteroids come in such a variety of types? Are these types thoroughly mixed throughout the asteroid belt? Can we relate these different types of asteroids to the meteorites in our museum collections? Do we know from which specific asteroid any meteorite comes from?

Some preliminary answers to these questions began to emerge from color measurements of the largest asteroids made a few years before. Asteroids located in what we call the main belt between Mars and Jupiter seem to have experienced a wide range of histories dating back to the formation of our Solar System 4.6 billion years ago. Our best understanding is that they are left-over building blocks and collision fragments of a planet that failed to form. The strong gravity of nearby Jupiter kept stirring the pot, causing the pieces to bump into each other with speeds that were too high to successfully stick together and form a single planet. The deep black carbon asteroids are objects that have never been subject to strong heating and these tend to be most common in the outer regions of the main belt. The reddish stony asteroids seemed to be planetary building blocks that were slightly heated, and these appear most commonly in the inner asteroid belt. The iron-rich asteroids also are more common in the inner asteroid belt and most likely are the iron cores of tiny planetary worlds (just a few hundred kilometers across) that were severely heated and melted, allowing the heavy iron to sink to the center. Collisions stripping away everything but the iron core have left these super-strong remnants. In a general way, similar classes of carbon-rich, stony, and iron-rich meteorites seem to correlate with these asteroid categories, but no specific links could be made.

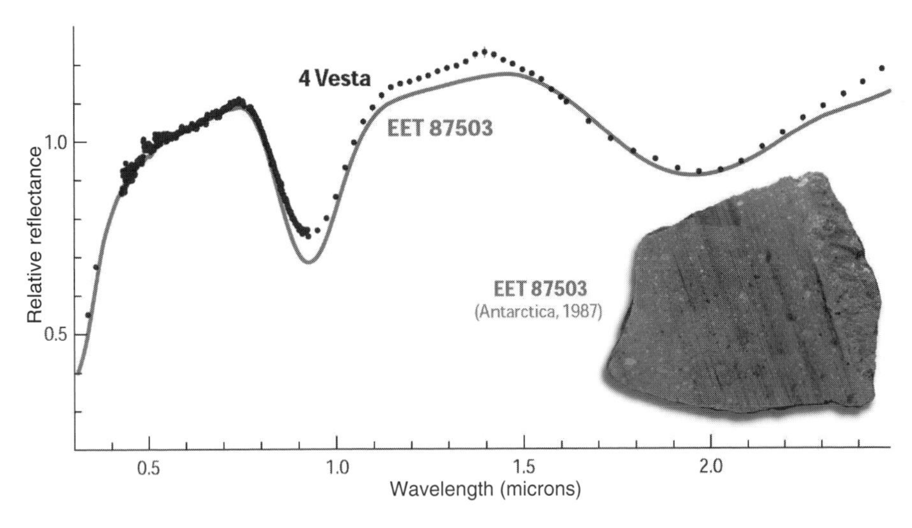

Figure 8.6 Comparison between the telescopic spectrum of Vesta and that of an HED meteorite. An HED meteorite is shown in an inset image. This figure appeared in *Sky & Telescope*, July 2001.

Little had I known that at about the same time when I was first looking at Vesta through my backyard telescope, professional astronomers in the 1970s were making their first detailed color measurements of this small world. Vesta's spectral colors looked nothing like the properties of any other known asteroid. Its unique signature indicated its surface was made almost purely of the mineral pyroxene. Such pure pyroxene is only known to exist in rocks that are formed from lava flows! Like asteroid B612 in Antoine de Saint-Exupery's Le Petit Prince (The Little Prince), it seems Vesta had a "volcano"! More correctly stated, Vesta somehow was so strongly heated that lava flows (probably just oozing from cracks rather than from a volcanic mountain) covered its surface. How amazingly appropriate that the name Vesta, denoting the goddess of the hearth (home fire), was bestowed nearly two centuries before the "volcanic" nature of Vesta was discovered.

The first measurements of Vesta's spectrum also raised an intriguing mystery with a seemingly impossible clue. The pyroxene-rich spectrum of Vesta was immediately recognized as a very close match to a class of basaltic (volcanic) meteorites known as HED's (for Howardites, Eucrites, and Diogenites). Since Vesta is the *only* large asteroid known to match these meteorites, was it safe to conclude that these meteorites were actual pieces of Vesta? If so, the 4.5 billion-year formation age of these meteorites reveals a fascinating history of the oldest volcanic processes in the Solar System. (More than 500 million years older than the lava plain or "seas" on the Moon.) Despite the circumstantial evidence, the prevailing answer was no, pieces of Vesta "can't get here from there" for at least two good reasons. First, Vesta's large size (500 kilometers in diameter) gave it sufficient gravity that it would be hard for cratering impacts to launch samples off of the surface and far into space. Second, Vesta's central location far from the edges of the asteroid belt and from any Kirkwood gap (regions that act as delivery services of asteroids and meteorites to the inner Solar System) made it highly unlikely that any piece of Vesta could ever be delivered to Earth.

As I began my own spectral studies of asteroids, Vesta and its decades-old meteorite conundrum were not particularly on my mind. I was on the hunt for something new among the asteroids. My goal was to learn whether the spectral colors of the smallest observable asteroids were similar to those of their larger counterparts and whether there were new clues to be found for matching asteroids to the complete variety of other meteorite types. Working with my graduate student, Shui Xu, we began collecting our data in Arizona and bringing it back to MIT for analysis. As we examined the data for hundreds of asteroids, the results for one needle in the haystack clearly stood out. The spectral colors for asteroid 3155 Lee (named after the Confederate general Robert E. Lee) showed an incredibly strong signature for the mineral pyroxene, just like Vesta! Most interesting was that the orbit of Lee placed it as a close neighbor of Vesta, and, in fact, it had been predicted to be a member of a possible Vesta family. Was it possible that Lee looked like Vesta because it was actually a huge chunk (7 kilometers across) blasted off of the surface by a giant cratering impact? Could this somehow help show that you can get meteorites from Vesta? Whatever the answers were, I was going to find them. Vesta was calling me back.

Building a bridge

It is a common tale that people get great ideas in the shower. For me, it has at least happened once. I was considering my Vesta problem while showering before heading to the airport to spend Thanksgiving 1992 at the MDM telescope in Arizona. (Yes, I have an incredibly patient and enduring wife and family.)

Over the course of a few months after discovering the Vesta-like spectrum of 3155 Lee, we measured dozens and dozens of additional asteroids throughout the main belt. Out of all of these, a total of ten were Vesta-like asteroids. Eight of these were in the vicinity of Vesta and were viable as Vesta family members. The other two were not at all close to Vesta's location. Instead, they were right on the edge of a relatively distant Kirkwood gap. Their spectral signatures looked just like the other eight, but they were so far away from the others they could not possibly be related. Or could they?

My shampoo-headed insight was that I could perform a test to see if they might be related after all. Using the parameters that describe the size, shape, and tilt of their orbits, I realized I could define a three-dimensional volume of space that would represent a "bridge" between Vesta and these two wayward objects. Performing my test required that I identify which, if any, known asteroids happened to occupy this volume. If some of these were also Vesta-like, then the wayward objects would no longer be isolated. There would be a clear trail of Vesta-like stepping stones between Vesta and the Kirkwood gap. Skipping the "delicious" airplane meals, I spent my entire flight time identifying known asteroids that we could observe as a test of my idea. I found five candidates. Being eager to know the results, Shui and I set up a data-processing system right at the telescope so that we could have a quick look at the spectral measurements as they were recorded.

One by one, we found and measured our five candidates. The first one? Vesta-like! The second one? Vesta-like! The third–fourth–fifth ones? *All* Vesta-like! Eureka! Not only were we dealing with pieces of Vesta in its immediate vicinity as a family, we were seeing evidence that 5–10 kilometer "Vesta-chips" had been flung far across the inner asteroid belt all the way to the Kirkwood gap. Through this "escape hatch," pieces of Vesta could reach the Earth. The Gordian knot was slashed – here was evidence that somehow Nature was launching samples off of Vesta on a trajectory that could reach the Earth. The first strong link between a specific group of meteorites and one specific asteroid world was forged! That world is Vesta.

The smoking gun

Even though we now had all of the evidence linking Vesta and its "chips" to the nearest Kirkwood gap for delivery to the inner Solar System, the case seemed almost too convincing. Surely the size of an impact that could have produced all of this debris would have destroyed Vesta. How could Vesta have survived an impact of this magnitude? Maybe Vesta had a twin sister living next door that was destroyed instead? Probably not and for good reason. The basalt (volcanic) layer is only a thin "icing" on the surface of the planet. For a Vesta-like body, this thickness may be only ten kilometers out of the 500 kilometers diameter. When destroyed, the basalt that gives the distinctive pyroxene-rich spectrum will be only a small fraction of the total debris volume. Instead the debris will be dominated by a different mineral, a greenish one called olivine that forms throughout a much thicker mantle layer more deeply below the surface. Olivine has its own distinctive spectral signature, and, because it is extremely rare or absent within the Vesta family, the matriarch of the family must still be intact. Impacts on to an intact Vesta-like parent would predominantly chip away at the pyroxene-rich basalt crust without frequently penetrating or excavating deeply into the olivine mantle. Because we see only the basalt surface layer mineral (pyroxene) dominating the spectral properties of the Vesta family, it seems certain that the progenitor must be intact. Vesta, being the only large basalt-covered asteroid known and being located right in the midst of a family of basaltic fragments, must surely be the parent.

While a scientifically sound argument makes a convincing case, cold hard evidence nails it shut. Like Sherlock Holmes, it was time to pull out the magnifying glass. Bring in the Hubble Space Telescope! In 1996, nearly five years after initially discovering the Vesta family, I joined with a team of colleagues to use the Hubble to image Vesta during a particularly favorable alignment with Earth when it was near its minimum distance of "only" 170 million kilometers. At that distance, Hubble could easily make out the size and shape of Vesta and provide some detail of its surface. We were filled with astonishment as we received and analyzed the images. There on the south pole of Vesta was a giant crater nearly spanning its entire (500 kilometers) diameter! Rising from the middle is a "central peak," commonly seen in large craters on the Moon, which results from the surface rebounding from the impact's plunge. With a volume more than 100 times that of all the Vesta family members combined, this giant crater was the evidence we were looking for. We had found the "smoking gun"! Clearly the Vesta family has been excavated from this site and we see the pieces of this family spread like stepping stones from Vesta to the Kirkwood gap for delivery to the vicinity of Earth. Thus not only can meteorites come from Vesta, they almost certainly do! I marvel at the thought that when I am holding one of these special (HED) meteorites in my hand, I am reaching out to touch the surface of Vesta.

Worlds within reach

Saturn is now in the midst of its second 29-year circuit around the sky since 1970, having recently passed by the position where I first spotted it. On a human time scale I can be hopeful to see Saturn pass by a second time, but very unlikely a third. The Cassini spacecraft will soon orbit and study Saturn in detail, immensely beyond my first backyard glimpse. It is enormously satisfying that in some small way my work on Pluto may contribute to a mission there becoming a successful reality. The intriguing volcanic nature of Vesta and the detailed planet-like history that its meteorite link conveys is leading NASA and international space agencies to move forward with plans for one or more missions to explore Vesta, including proposals of my own. My grandfather (now honored with the name of my first discovered asteroid 13014 Hasslacher) was right: These worlds are within reach to those who choose to strive for them.

Further reading

Binzel, R. P. (1990). Pluto, *Scientific American*, June.

Binzel, R. P. (2001). A New Century for Asteroids, *Sky Telescope*, July.

9

Titan: a moon with an atmosphere

Christopher P. McKay, NASA Ames Research Center

Chris McKay is amazing. A native of California (though he grew up in Florida and went to grad school in Colorado), he simply must be either a magician or must have cloned himself, for he accomplishes far too much for any one human being. Chris is an expert on practically every topic in planetary science, a prolific writer, an explorer of mountains and arctic regions, an accomplished diver, and a husband and father. Chris's favorite research topics are Mars and Saturn's moon Titan, each of which he would like to visit. If anyone could, it would be Chris.

Almost twenty years ago my interest in planets and life led me to Titan: the largest moon of the planet Saturn and the only moon in our Solar System with an atmosphere. In a real sense all the work I have done over the past two decades on Titan has been to prepare for a two-hour descent through the atmosphere of that world by the Huygens spacecraft now scheduled for January 2005. Why, you might wonder, would I be so interested in a place that I would spend 20 years working on it knowing that I will never visit it myself and that the only spacecraft likely to reach it in my lifetime will only return data for a mere two hours? Let me answer this question by showing you that this is a world like no other, and yet a world that in several strange ways is like our own Earth.

What interests me the most about Titan is its atmosphere and organic haze. Venus and Mars are often described as our sister planets but, in fact, Titan's atmosphere is the most similar to that of the Earth. Venus and Mars have atmospheres composed mostly of carbon dioxide (CO_2) and the pressure on these worlds is 100 times higher (Venus) and 100 times lower (Mars) than Earth's pressure. Titan's atmosphere, like the Earth's, is mostly nitrogen (N_2), and the pressure on Titan is only 1.5 times higher than sea-level pressure on Earth. In addition to N_2, Titan's atmosphere contains a few percent of methane (CH_4).

Photochemical reactions triggered by sunlight result in the formation of complex organic molecules in the atmosphere of Titan. Ultimately these organic molecules coagulate into solid organic particles and form a thick haze in Titan's atmosphere. The surface temperature of Titan is a chilly −180 degrees C, much too cold for water to be present on the surface as liquid or for water vapor to be in the atmosphere. However, liquid CH_4 is probably present on Titan and may form clouds in the lower atmosphere. In many respects CH_4 on Titan may play the role that water does on Earth.

In this chapter I will take you on a personal tour of Titan. The tour begins with the atmosphere and its interesting greenhouse and anti-greenhouse effects, and how they compare with Earth. Then we consider the organic chemistry of the atmosphere and wonder about CH_4 clouds and seas of liquid CH_4 and ethane (C_2H_6). Then we see how our understanding of Titan's present state can help us understand how it might have been in the distant past and how it might be in the distant future.

We have learned a great deal about Titan from the Voyager spacecraft that flew by that world two decades ago. But much more will be revealed when the Huygens probe flies through the atmosphere and lands on the surface of Titan in 2005.

A greenhouse effect with a difference

Titan's greenhouse effect is second only to Venus in strength: 92% of the energy that warms Titan's surface comes from its greenhouse effect. Venus with its thick atmosphere has the strongest greenhouse of all the planets. Its surface is at a temperature of 460 degrees C, and 99.9% of the energy that warms the surface of that planet comes from its greenhouse effect. The greenhouse effect does not create energy but it can be an effective trap and recycle the energy from the Sun.

However the gases that cause the greenhouse effect on Venus, Earth, and Mars are CO_2 and water (H_2O), and neither are abundant in Titan's atmosphere. Titan's greenhouse effect comes from a very different set of molecules: N_2, CH_4, and hydrogen (H_2). It took us many years to understand the greenhouse effect of Titan and how it compares to Earth. This was the first problem I worked on when I began to study

Titan and it remains my favorite aspect of Titan's atmosphere. I began this work in 1984 at NASA Ames as a young scientist under the direction of James Pollack. Jim was Carl Sagan's first graduate student and he and Sagan had conducted the first studies into the greenhouse effect on Venus, Mars, and Earth. I was working with the world's expert on the topic. With us was a young French scientist from the Paris Observatory, Regis Courtin, who was visiting NASA for a year. We would meet regularly at the table in Jim's office. Since Jim passed away I am now in that office and I write these words at the same table where Jim, Regis, and I used to have our meetings.

An atmosphere creates a greenhouse effect when it is transparent to sunlight but blocks the escape of thermal radiation from the surface. On Earth, CO_2 and H_2O vapor in the atmosphere create the greenhouse effect. Without this warming effect created by the atmosphere the average temperature of the Earth would be −15 degrees C rather than +15 degrees C! As a result, energy can come in easily but cannot leave easily. Human activities on Earth are increasing the CO_2, and adding other strong greenhouse gases to the atmosphere. As a result, the temperature of the Earth is increasing, and this increase in temperature causes an increase in the amount of water vapor in the atmosphere. Since water vapor is also a potent greenhouse gas, this results in a positive feedback, further warming the Earth. It is clear that human activities are augmenting the greenhouse effect and that the feedback from water vapor is amplifying that effect. What is not clear is the magnitude of the human-induced effect and how it compares to natural variations in temperature.

As I mentioned above, Titan's greenhouse effect is due to CH_4 and H_2, and the first puzzle we encountered was how could these gases cause a greenhouse effect at all? The conventional wisdom is that symmetrical molecules like these do not cause greenhouse effects because a gas molecule can absorb thermal radiation only if it is asymmetrical. Earth's atmosphere is mostly N_2 and it creates no greenhouse effect on this planet. However on Titan the atmospheric density is much higher than on Earth due to the low temperatures and the slightly higher pressures. The N_2 and other molecules on Titan are packed in five times denser than the air on Earth. At this higher density the molecules

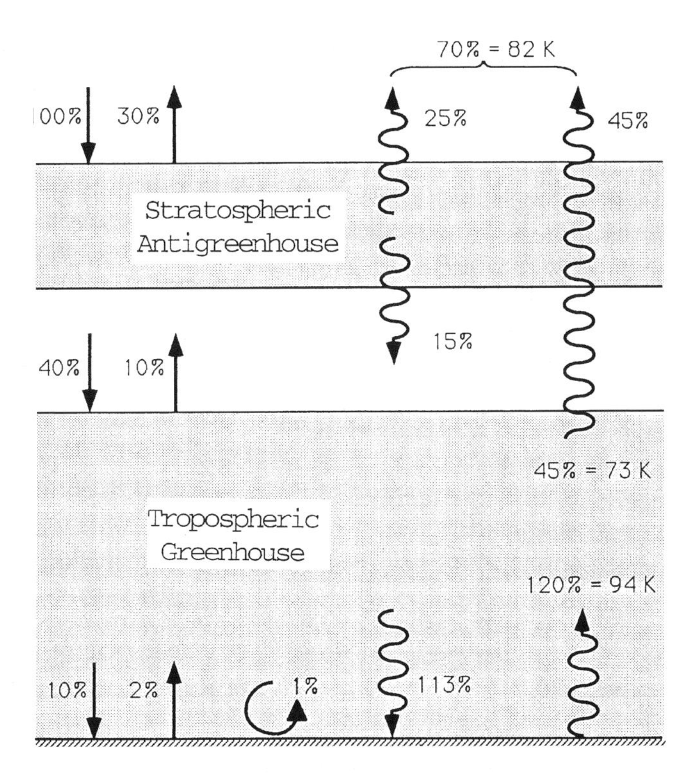

Figure 9.1

The greenhouse and anti-greenhouse effects in Titan's atmosphere are shown here. The greenhouse effect arises when an atmosphere is transparent to sunlight (straight arrows) allowing sunlight to warm the surface. (A small amount of the surface energy on Titan is carried away by convection, 1%, shown as the circular line.) The warm surface glows with infrared radiation (wavy arrows) but the atmosphere blocks the escape of this radiation and directs some back to the surface. On Earth, Venus, and Mars, the greenhouse gases are CO_2 and H_2O, while on Titan they are N_2, CH_4, and H_2. Titan has the second strongest greenhouse effect in the Solar System, after Venus. The anti-greenhouse effect is caused when a layer in the atmosphere blocks sunlight but is transparent to infrared radiation. This is just the opposite of the greenhouse effect. Titan's organic haze produces a strong anti-greenhouse effect, the only one in the Solar System.

bounce into each other and while they are rebounding they are distorted. In this distorted, asymmetrical state they can absorb thermal radiation and can thus cause a greenhouse effect similar to the naturally asymmetrical molecules of CO_2 and H_2O.

Fortunately for our work, Regis Courtin was working on a method to compute the effect of collisions in causing a greenhouse for Jupiter and Saturn. With a little effort he was able to adapt his method to Titan and we had the missing piece needed to do a complete study of the greenhouse effect on Titan. We put together a sophisticated computer program that calculated the temperature of Titan's atmosphere based on the sunlight coming in through the organic haze and the greenhouse effect of the atmosphere.

A few years earlier, 1980, the Voyager spacecraft had flown past Titan. As the spacecraft flew behind Titan it sent its radio beam through Titan's atmosphere. The beam was bent by the atmosphere; from the degree of bending the temperature profile of the atmosphere can be determined. We compared our computer-simulated results to these measurements. Overall our results were in good agreement (once we included the anti-greenhouse effect discussed below) but only when we doubled the greenhouse effect of CH_4 compared to what we expected it to be. At the time we did not worry too much about this since the uncertainties in the theory for the CH_4 greenhouse effect were large enough that a factor of two could fit. However, a few years later this little factor of two would prove to be an important clue to one of the most unusual features of CH_4 in Titan's atmosphere.

The anti-greenhouse effect discovered

As we worked to understand Titan's greenhouse effect we realized that another effect was present. While the greenhouse gases in the lower atmosphere were warming Titan, the organic haze in the upper atmosphere was having the opposite effect. Titan, in addition to having a greenhouse effect had what we eventually came to call an "anti-greenhouse" effect. The greenhouse is caused when there are gases in the atmosphere that are clear to sunlight but opaque to thermal radiation. Suppose that an atmosphere had a layer in it with just the opposite properties: it was opaque to sunlight but allowed thermal radiation to pass right through. Such a layer would be an anti-greenhouse layer and it would cause a cooling rather than a heating.

There was a lot of interest in anti-greenhouse layers because just a few years earlier, in 1982, Jim Pollack, Carl Sagan, and others had considered them as one possible result from a major nuclear war. They called the resulting cooling "nuclear winter." Even earlier, when I was graduate student at the University of Colorado, I had spent the summer of 1980 working on a team led by Jim Pollack looking at how an anti-greenhouse layer might have formed after a giant meteor impact 65 million years ago. We concluded that the darkness and cooling due to an anti-greenhouse layer may have been the main cause of the dinosaur extinctions. Having only considered the anti-greenhouse effect in these hypothetical – and decidedly undesirable – cases, it was interesting to realize that there was a real and very strong anti-greenhouse layer in Titan's upper atmosphere.

Titan's anti-greenhouse layer is caused by an organic haze produced by sunlight reacting with CH_4 and N_2 in the upper atmosphere. The haze is very similar to urban smog. The organic haze particles are dark to sunlight and block almost a third of the sunlight reaching Titan. However the thermal radiation from Titan's surface passes right through these particles. Our computer simulations showed that the anti-greenhouse effect on Titan was 50% as strong as Titan's greenhouse effect, so the net effect was still a greenhouse. Titan has the second strongest greenhouse effect in the Solar System, but it has the strongest anti-greenhouse effect!

Organic chemistry and Titan in a jar

When we were at the start of our studies of Titan's anti-greenhouse smog, we wondered about the type of organic material making up the smog. We needed this information in order to be able to compute the effect of the smog on sunlight and thermal radiation. Coincidentally, a research group at Cornell University was working to synthesize in the laboratory an organic smog like that found on Titan. Carl Sagan, Bishun Khare, and their co-workers at Cornell University, put a mixture of gases in a jar made up of the same compounds as Titan's atmosphere. They then sent an electrical discharge through the mixture. After a few hours a brownish goo formed on the wall of the jar. Sagan called this organic material made in the laboratory "tholin," after the Greek word for mud. Working with the group at Cornell we began a long study into the nature of the Titan tholin, trying to see if it really was the same as the material we could see on Titan.

The first test was to compare the color of Titan with the color of the laboratory tholin. Telescopes from Earth have been able to get very precise measurements of the brightness of Titan from the ultraviolet to the far infrared. In the ultraviolet Titan is very dark: it would look as black as coal. However, in the red and the infrared Titan is bright: as bright as freshly fallen snow. We were able to show that the laboratory tholin had exactly the right color to match this pattern of color seen on Titan. In addition, other materials made in the laboratory, but not from Titan mixtures of gases, did not match the colors of Titan. Thus we were pretty sure that the stuff made in the Titan jar was very much like the stuff being made on the real Titan.

Now we had a recipe for making Titan's organic haze in our laboratories. So, all sorts of experiments came to mind. The most interesting one was: Would the organic material on Titan dissolve into the liquid CH_4 and C_2H_6 that is likely to be present on the surface as lakes or seas? If it did dissolve and form an interesting organic soup perhaps this could be the basis of a new type of biology. Knowing how to make Titan tholin allowed us to begin direct experiments about possible life, based on liquid CH_4 and C_2H_6.

Joan Mathog was a high school student in San Jose, California, who was working with us as a student intern at the time. She was very interested in this question and decided to do the experiments as a science project. She prepared a glass jar and set up a tank of CH_4 and N_2 so that these gases flowed through the jar in such a way that the mixture in the jar was the same as Titan's atmosphere. She then arranged three spark generators to alternately spark the mixture. We ran the sparks for several weeks, and a thick layer of Titan tholin built up on the walls of the jar. Joan then carefully opened the jar and scraped out the tholin. Joan had built a Titan tholin-making machine. I think we made more Titan tholin in that experiment than had ever been made before – or since. The next step was to cool some C_2H_6 down until it became liquid. We used liquid nitrogen for this. Then we stirred the tholin into the C_2H_6, and poured the mixture through a fine filter. Only material dissolved in the liquid would pass through the filter. Tests with water showed that our approach would work; it also showed that water would be a great solvent for Titan's organic chemistry. Unfortunately for biology Titan is much too cold for liquid water to exist. Also unfortunately for biology we found that virtually nothing dissolved in the cold liquid C_2H_6. The filtered liquid was just as clear as the pure C_2H_6. And when we evaporated it, there was no residue left over. It would have been fun to discover that there was an interesting organic chemistry in Titan's liquids. But in science a negative result is important too. Joan presented her results at an international scientific meeting in Berkeley California in 1986, and eventually these results were published in a paper reviewing the properties of Titan's organic chemistry.

Send in the clouds

Jim, Regis, and I continued to work on refining our calculations of Titan's greenhouse and anti-greenhouse effect. One aspect that stuck in the back of my mind was the doubling of the greenhouse effect due to CH_4. I wondered if more precise data and simulations would show that this doubling was real. A few years later Regis had put together a much-improved calculation for the greenhouse effect of CH_4 that was based on new laboratory data and more complete calculations. These new results were precise enough; the uncertainty was much less than a factor of two. Furthermore, these results showed that the factor of two was not acceptable; the CH_4 greenhouse was half as strong as we needed to explain Titan's greenhouse effect. It was a real puzzle. I worked for months fiddling with the computer code trying to make the greenhouse effect work and predict the correct surface temperature for Titan. It would not happen; the CH_4 greenhouse effect was not strong enough. Then on a trip to visit Regis in Paris, he suggested a brilliant solution: What if the greenhouse effect of CH_4 is correct but there is twice as much CH_4 as we think there is? We realized immediately that this would do it. But how could there be twice as much CH_4 as we expected? We thought we knew how much CH_4 there was on Titan.

The amount of CH_4 on Titan should be limited by its condensation to form clouds. If there was more CH_4 in the atmosphere it should just form clouds and rain out. It was not clear how Titan could have twice as much as the limit set by cloud formation. I was in France for a week and Regis and I worked that week on the problem and came up with a possible answer: Maybe CH_4 was not forming clouds on Titan. Perhaps for some reason cloud formation was inhibited. This happens on Earth with water and ice, but not very often, since sea salt is a prevalent particulate and makes excellent seeds for water clouds. One possible reason it could be happening with CH_4 on Titan is the absence of suitable small particles for the clouds to form on. These "seed" particles are crucial to cloud formation. Joan's experiments had shown that Titan's organic material was not good seeds for CH_4 clouds because they were not soluble in C_2H_6, and by extension CH_4.

Using a theory for how CH_4 forms clouds, Regis and I were able to show that if cloud formation was suppressed, due to the insolubility of the organic seeds, then CH_4 in Titan's atmosphere could be twice the level expected. Coincidentally, Robert Samuelson, at NASA's Goddard Space Flight Center in Maryland, came up with the same idea at about the same time, independently. We could compare notes because Bob was visiting France also and our visits would in fact overlap. The fact that his results agreed with ours gave us (and him) confidence in publishing this novel hypothesis.

One of the nice things about this work was that this was a hypothesis that could be tested. If cloud formation on Titan was inhibited, then we would not expect solid cloud decks on that planet the way clouds are present on Venus or Earth. Instead we would expect no clouds, or at most transient clouds. We do not yet know the state of clouds on Titan but some preliminary observations by Caitlin Griffith, of Arizona Northern University, suggest that clouds may be patchy and transient. When the Huygens probe goes through Titan's atmosphere it will directly measure the CH_4 concentration and it will look for clouds. I will be standing by interested to see if our hypothesis of double CH_4 is correct.

Titan's early days

In his book, *Life on the Mississippi* (1884), Mark Twain observed that, "There is something fascinating about science. One gets such wholesale returns of conjecture out of such a trifling investment of fact." Indeed, I have always enjoyed this aspect of science and could not resist using our computer simulation of Titan's present atmosphere to conjecture about what conditions were like on Titan billions of years ago and what they might be like billions of years in the future. We added to our team Jonathan Lunine, of the University of Arizona, who had worked extensively on Titan's early history.

Starting with the past: As Titan formed we know it must have been hot due to heating released by the in-falling material. To simulate this, we told the computer to add heat flow to the bottom of the atmosphere. Because Titan's atmosphere is so thick it did not take much heat flow before Titan's seas and lakes evaporated into a runaway greenhouse. We concluded that as Titan formed it would have been even more like Venus than it is today.

But after a hot start our simulations predicted that Titan would have quickly become very cold. Once the gravitational energy that was released as it formed was gone, Titan would have only been warmed by the Sun. But the Sun was significantly dimmer four billion years ago. In fact, astrophysics compute that it was only 70% as bright as it is today. When we turned the Sun down that much in our computer simulation, Titan froze up. The CH_4 and N_2 in its atmosphere would have condensed completely on the surface to form ice. Titan would have looked like Triton, the moon of Neptune, looks today: a frozen ice ball. Over time as the Sun became brighter our simulation predicts that Titan became warmer. About half a billion years ago we expect that Titan's atmosphere became thick like it is today.

The Sun will continue to get brighter and our simulations predict that Titan will get warmer still. In less than a billion years it will again get so hot that any CH_4 seas or oceans on the surface will evaporate and it will have a runaway greenhouse. Once that happens the surface may become warm enough that the water that is currently frozen solid on the surface might melt and Titan's chemistry would change drastically. Simulating the interaction of the liquid water with the CH_4, N_2, and organic material is an interesting challenge . . . perhaps I will work on this. This future epoch could be Titan's best chance at having life, at least life as we know it based on liquid water and organic molecules.

Waiting for Huygens

We have never visited Titan; we have only observed it from Earth and from spacecraft flying by. This will change in 2005 when the Huygens probe will enter into Titan's atmosphere and parachute down and land on the surface. For slightly more than two hours we will get detailed data on the nature of the atmosphere, the presence or absence of clouds, the composition of the organic haze, and photographs of the surface. If past missions are any guide we can expect some of our current hypothesis to be confirmed by the probe data, others to be proven wrong, and best of all many new puzzles revealed about this mysterious world with its Earth-like atmosphere.

Stay tuned. I will.

Further reading

Coustenis, A. and F. Taylor (1999). *Titan: The Earth-like Moon*, World Scientific.

Coustenis, A. and R. D. Lorenz (1999). In *Encyclopedia of the Solar System* (P. Weissman, L. McFadden, and T. Johnson, eds.), San Diego: Academic Press, pp. 377–404.

McKay, C. P., J. B. Pollack, and R. Courtin (1992). Titan's Greenhouse and Anti-greenhouse Effects, *The Planetary Report*, **12**, May/June.

Owen, T. (1982) Titan, *Scientific American*, **246**, February: 98–101, 104–109.

10

Seasons at the edge of night

Alan Stern, Southwest Research Institute

Alan Stern is the father of three children, a husband, a pilot, a scientist, and an author. He grew up in New Orleans and Dallas, and has lived most of his professional life in Colorado. Alan's research interests focus on the formation and evolution of the outer solar system, most particularly Pluto, comets, and the Kuiper Belt. He is a passionate advocate of space exploration – both human and robotic, and he is the director of the Space Studies Department of the Southwest Research Institute. He is the Principal Investigator of NASA's Pluto-Kuiper Belt mission, called New Horizons. Alan enjoys gardening, writing, hiking, and camping.

I did not really intend to become an astronomer, though I had loved the subject since I was seven or eight. And I certainly never thought much about Pluto as a boy. I wanted to be an astronaut. And, by 1980, I believed I was well on my way. I was 23, and I had twin bachelor's degree in physics and astronomy, and a master's degree in aerospace engineering. I was already a commercial pilot, a skydiver, and a scuba diver. People said the odds were long, but the world was young, so why not follow your dream?

One part of my strategy to become an astronaut was to prove myself a good generalist, that is, proficient in multiple technical fields. After all, being versatile is one of the hallmarks of a good astronaut. This in mind, I elected to avoid the specialization of a Ph.D. and instead pursued a more general course, by earning a second master's degree, this time in atmospheric science.

I never did become an astronaut, though. I interviewed as a mission specialist finalist in 1995 and I was recommended to fly as a payload specialist about a year later. But that is another story. So is the one about finally getting that Ph.D. I'd opted out of in the early 1980s.

The story I'll tell here is about Pluto, which I adopted as a thesis topic for that master's degree in atmospheric physics, and which has fascinated me ever since. In what follows, I will tell you a bit about how I came to be interested in Pluto. I will also describe the way we came to coin a much used term regarding Pluto, "atmospheric collapse." And in doing so, I will describe what we know about Pluto's atmosphere. Let us begin.

Beginnings

I took my undergraduate and master's training at the University of Texas in Austin, the administrative base of McDonald Observatory. I lived in Austin from 1975 to 1981, and in that time I earned four degrees. In 1980, after taking my graduate class work in the atmospheric sciences (Texas had no planetary atmosphere program), I knew that I would also have to write a research thesis. Being an inveterate space person, and loving solar system astronomy (which specializes in places you can actually go to), I decided to try to combine my astronomical background and interests with the atmospheric sciences.

The planetary science group at the University of Texas was not very large in 1980 (it is not now either), but it did have some very good people. I knew a few of them casually from my undergraduate days, when I had regularly been employed as a grader and a research assistant in other crannies of the astronomy department. One person I did not know, but knew about, was Larry Trafton. Trafton was a hotshot young astronomer in his late thirties, with outstanding training from Caltech and the Air Force, which he exploited to make a variety of important discoveries regarding the atmospheres of the moons of the giant planets. Trafton was also, like myself, a pilot and was interested in flying in space. I had heard he was tough on students, but I was interested in studying Pluto, and Trafton had just completed a landmark paper about the escape of Pluto's suspected atmosphere (this paper later turned out to be wrong, but it nonetheless got a lot of people thinking about Pluto's possible atmosphere).

Why was I interested in Pluto? Because it seemed so *wide open* as a research topic. So little was known about this world beyond the giant planets that I concluded that almost anything I could contribute would be something new and useful. As a neophyte researcher, that was an attractive prospect indeed. And given that I was about a year from a deadline my chemical executive father had imposed ("finish up, four degrees is enough – get a job"), it seemed Pluto was an easy mark – a subject I could master, make some quick strike research results, and finish school. It never crossed my mind that the more I would learn about this little world so far away, the more I would want to learn about it.

Ignorant but determined, one day in the summer of 1980, I introduced myself to Larry Trafton and told him I wanted to work on Pluto. He said he did not have any money for a student (translation: Go away). I responded that I was teaching flying so I did not need any money – I would work for free. Larry, an avid pilot himself, immediately perked up. We talked about flying for a while. Then we talked about Pluto. By the end of the meeting, Larry sent me off to read some papers and to start a weekly dialog about what useful, master's level research topic we might address.

Dark ages

So I read about Pluto, and I also read about the atmospheres of Saturn's moon Titan and Jupiter's moon Io. Owing to their lower gravity and their location in the cold outer solar system, it was suspected that Pluto's atmosphere, if it had one, would have more in common with satellite atmospheres than the atmospheres of the giant planets, Earth, Mars, or Venus. At the time, Titan and Io were the only satellites with confirmed atmospheres, so I expected to draw physical insight from what was known about these loosely kindred bodies.

In 1980, our knowledge about Pluto was pretty meager. Why? Primarily because: (i) the technology of the day was too primitive to study a planet so faint and far away, and (ii) no space mission had ever been there.

In brief, what had been eked out in the 50 years since Pluto's discovery (that is, from 1930 to 1980) was this: Pluto was known to orbit the Sun in 248 years along an elliptical path that took it from 30 to 50 AU (AU, or astronomical unit, is the distance from the Earth to the Sun). No other planet follows such an elliptical orbit.

Figure 10.1

Pluto and Charon, far against the deep as imaged by our Hubble Space telescope team (Alan Stern, Marc Buie, and Larry Trafton) in 1995. (pluto_gem.ps)

In 1978, astronomers had discovered Pluto has a satellite, which was subsequently named Charon. Pluto's rotation axis was known to be tilted rakishly, some 120 degrees, relative to its orbit plane. This extreme tilt is reminiscent of Uranus's 98-degree polar tilt. The length of Pluto's day had been measured by monitoring the repeating rise and fall in its brightness every 6.387 days. As it turned out, this was also found to be the orbital period of Charon, indicating that the satellite was in synchronous orbit

Pluto's size was not known in 1980, but there were some constraints on it. Pluto's radius was at that time estimated as between 900 and 2,500 kilometers (560 and 1870 miles). We also knew Charon was fairly large in comparison to Pluto. In fact, in 1980 Charon's size was determined by timing its passage in front of a star. From this measurement it was learned that Charon's radius is 600 kilometers (375 miles), give or take a few percent. (I believe this is the first and only time the size of a planet's satellite was known before the size of the planet itself was known.)

From Charon's orbit it had been possible to determine Pluto's mass, about 2/1,000ths that of the Earth. But because Pluto's radius was so uncertain, we did not know Pluto's density, and so we could not determine if it was primarily made of rock or ice. However, based on Pluto's estimated mass and its crudely constrained size, it was clear that the surface gravity was somewhere between a few percent to at most 20 percent that on Earth.

Pluto's color was known to be slightly reddish – less so than Mars but more so than the Moon. From spectroscopy, it had been learned that frozen methane is present on Pluto's surface. Because frozen methane looks a lot like frozen water snow from a distance, it was suspected that Pluto's surface was highly reflective; however, this could not be quantified because we did not know Pluto's radius, and therefore could not derive an albedo (reflectivity) from the planet's brightness.

There was no direct evidence for an atmosphere on Pluto in 1980. The astronomical spectroscopes of the day were simply not able to see it (they now are). We knew that if a star were to pass behind Pluto, a rare event given Pluto's small size (about 1/2,000,000th of a degree across), we could check for an atmosphere by studying the way that the starlight diminished, but no such occultation was in sight (the first one observed would not come until 1985).

Nevertheless, we could surmise that Pluto had an atmosphere. The logic was as follows. Based on Pluto's distance from the Sun and the circumstantial evidence for a high surface reflectivity, we could estimate that the surface temperature was probably in the range 40 K to 70 K. Given this, and the well-known thermophysical properties of methane ice (which had been studied by physicists, often working for oil companies), we could calculate how much ice would sublimate (turn into gas) as a function of temperature. As it turned out, this kind of model calculation indicated that if Pluto's atmosphere was dominated by methane, it might have an atmospheric pressure as high as 1/10,000th that of the Earth's atmosphere if the temperature was near 70 K. However, because the pressure prediction depends sensitively on the assumed temperature, a 40 K surface would mean that the atmospheric pressure was far lower – about 1/10,000,000th that of Earth's atmosphere. Quite a range of uncertainty, surely, but those of us interested in modeling the atmosphere knew it was very likely that Pluto's atmosphere would one day be discovered. Based on this, it was open season on making models to predict what that atmosphere might be like.

Collapse

When Larry Trafton and I began our work on Pluto, one of the first things we wanted to explore was the question of what other major constituents might make up Pluto's atmosphere. We knew that the extremely low surface temperatures on Pluto severely limited the list of potential constituents. Many common gases, like carbon dioxide (prominent in the atmospheres of Venus and Mars) and water vapor (common in the atmosphere of Earth and the giant planets), would be frozen out, and thus unable to exist as a gas at Pluto. Other gases, like hydrogen and helium, though able to exist as a gas at Pluto's cryogenic temperature, were so light that they could escape Pluto's feeble gravity.

There was however, a short list of gases that were able to fit a three-part criteria that we laid out which would make them possible major constituents of Pluto's atmosphere. To make our list, a molecular species had to: (a) be able to exist in gas form at temperatures of 40 to 60 K, (b) be heavy enough not to escape from Pluto so rapidly that it would have been lost over time, and (c) be abundant in the solar system (there wasn't much use in exotic man-made cryogenic molecules that wouldn't be expected to occur in nature). I spent a lot of time in the astronomy department's small technical library ("the Perdier collection") poring over tables of numbers to sieve through all of the gases known to occur in the atmospheres of the solar system to see what species might make our Pluto list.

Six possibilities rose to the fore. These were: molecular nitrogen (N_2) and molecular oxygen (O_2), the noble gases neon (Ne) and argon (Ar), carbon monoxide (CO), and methane (CH_4). Methane, of course, was already strongly suspected to be in Pluto's atmosphere because it had been detected as a frost on Pluto's surface. Ne and Ar are noble gases that were known to have been relatively common in the solar nebula;

additionally, Ar is created in planetary interiors by the radioactive decay of potassium bearing rocks. CO is common in the atmospheres of Mars and Venus. N_2 and O_2 were the most interesting to me because they are the very stuff we breathe in every breath – how exotic to think the same stuff might make up Pluto's atmosphere!

We didn't know which of these species were most likely to actually be in Pluto's atmosphere (other than methane), but, nonetheless, we could examine the implications of the various cases.

One of the first things we looked at was the surface pressure that each of these molecules would produce in an atmosphere created by the sublimation (that is, the vaporization) of surface frosts by sunlight. This pressure, called the "equilibrium vapor pressure," could be calculated by knowing only the surface temperature and the abundance of each species on the surface. Since we did not know the species abundances, we simply made the assumption that each species made up 100% of the surface. This "limiting case" assumption was not necessarily realistic, but it did provide an idealized case we could work with in the absence of real compositional constraints; and it also provided an upper limit to the expected pressure of each possible atmosphere.

Of course, the other thing that we did not know was the surface temperature. As I pointed out above, the equilibrium vapor pressure of each species, and therefore the expected abundance of each species, is exponentially sensitive to the surface temperature. So there would be a large difference in the gas abundance one would predict if the temperature was 45 K or 50 K or 60 K. I calculated the abundance for each species on a handheld HP calculator in the library one afternoon. The numbers I got are shown in Table 10.1 (the units are the number of molecules from the surface to space, in every square centimetre of atmospheric column, which were calculated making the simplifying assumption that the atmosphere and the surface have the same pressure).

As you can see, there was an enormous range of possibilities. But despite this, there were also some real implications we could draw from these simple calculations. First, we could rule out the really extreme cases, like a neon atmosphere at temperatures above 40 K or so, because such an atmosphere would be so dense that its presence would be revealed by Raleigh scattering, making the planet appear blue, like

Table 10.1.

Species	Perihelion abundance (58 K)	Aphelion abundance (45 K)	Ratio
N_2	2×10^{25}	4×10^{23}	52
O_2	2×10^{24}	2×10^{22}	110
CO	7×10^{24}	6×10^{22}	125
CH_4	7×10^{22}	3×10^{20}	262
Ne	6×10^{29}	2×10^{28}	3
Ar	1.5×10^{24}	1.5×10^{22}	100

Earth and Neptune. Because Pluto was known to be red, not blue, we could eliminate such cases. Second, it was clear that the methane that was so likely to be in the atmosphere did not have to be the main constituent – it might even just be a trace, dwarfed in abundance by nitrogen or another similarly volatile species, if those species were abundant on Pluto's surface (something we later learned was in fact the case).

Most importantly, however, we concluded that, because the temperature dependence of the pressure is so steep, Pluto's elliptical orbit would be likely to generate wild swings in atmospheric pressure and density with distance from the Sun.

What a fantastic thought! When Pluto is close to the Sun, it might have a bulky atmosphere, and when it is far from the Sun, it might have very little atmosphere at all. In this regard, I recall imagining that Pluto might act like a comet, but on a planetary scale.

Now this was something – here we could hope to see every 248-year cycle an atmosphere grow from nothing to a real atmosphere, replete (we calculated) with winds, weather, day–night cycles, and even chemistry, and then ebb again, to be reborn the orbital next cycle! And to top it off, each of the candidate species had a different vapor pressure curve (and hence a different ratio of minimum to maximum pressure). This meant that the relative proportions of the various gases could vary over each orbital cycle, so that the dominant species in the atmosphere could vary with orbital location too. What a wonderland!

Just imagine, when Pluto was last at its farthest point from the Sun, aphelion, back in the 1860s during the American Civil War, Pluto may have had little or no atmosphere by comparison to today.

Its "atmosphere" would be frozen on its surface as a snow. Then, as Pluto came nearer to the Sun, and warmed, the snows would sublimate again into gas, making an atmosphere appear, perhaps as recently as the 1960s or 1970s.

Larry and I knew in 1980 and 1981 when we did this work that Pluto was then approaching its closest point to the Sun, or perihelion. Pluto would pass perihelion in 1989. We could expect to see the atmosphere reach a maximum about then, or perhaps in the decade or so to follow (as the lag between the change in heating and an atmospheric response could be calculated). Then, as Pluto moved away from the Sun, we reasoned, the heating would drop. The temperature didn't need to drop fast, or even very far, for the atmosphere to begin to decline. For gases like CO, N_2, and CH_4 , every 2 degree drop in temperature would decrease the atmospheric pressure by half. And that wasn't all: because the snow condensing out of the atmosphere on to the surface would brighten the surface, the cooling would be accelerated by the snow. That is, the loss of atmosphere would feed on itself in what is commonly referred to in science and engineering as the "closed loop feedback" effect.

We set out to model this phenomenon on a computer. At the time, that meant hand-writing a code, typing it on to *paper* punch cards (one card per line of code), and feeding them into a "mainframe" computer that the campus kept in a huge room with blinking lights over at "central computing." We wrote the code in Fortran. Debugging was a real joy. If the run did not work, either due to a coding error, a typo on a card, or something else, one had to deduce where the problem was (this sometimes took several more trial runs), write new code, type and insert the new punched cards into the stack (in the right place, be careful!), and begin again. It *was* fun, but I do not wish for those days again. I will take the PC-based computing and terminal typing environment of 2001, replete with Microsoft's own vagaries, over the mainframe and punched card days of 1981 – every time.

After I wrote and debugged the code, I ran it for many combinations of cases: different gas species from the table above, differing maximum temperatures, differing snow reflectivities, etc.

What we found was that Pluto's atmosphere would very rapidly snow out on to the surface, sometime between the mid 1990s and perhaps

2025. We couldn't be precise because there were so many variables, including, most importantly, what the main constituent of the atmosphere even was. But what we could tell was that the process could happen very quickly. Indeed, we estimated that it might take no more than a decade (less than 4% of Pluto's orbit) for the pressure of the atmosphere to decrease by a factor of ten from its maximum; indeed, we found that in some runs, the process could go up to 100 times faster – a factor of ten decline in just months! And it did not need to stop with a factor of ten decrease: the atmosphere could reach a minimum state as deep as several hundred times less dense than at its maximum! This effect was so precipitous that we gave it a name: *atmospheric collapse*. Our results were published in my master's thesis (1981), and then in a refereed journal paper in 1984.

Fast forward

Twenty years have passed since the initial atmospheric collapse work Larry Trafton and I did. Since then, Pluto's radius (1175 ± 25 km, or 734 ± 34 miles) was determined from both its eclipses with Charon that occurred in the late 1980s, and a stellar occultation that occurred during the same period. And Pluto's atmosphere was detected via stellar occultations that occurred 1985 and 1988. The precise surface pressure is uncertain (because we still do not know the *exact* radius of the planet), but the surface pressure is likely to be between 3 and 30 millionths of the pressure on Earth at sea level. The 1985 and 1988 occultations also provided some evidence for hazes and (or) complex temperature variations in Pluto's atmosphere. Later, methane was detected in Pluto's atmosphere by very sensitive, high-resolution spectroscopy in the mid 1990s. N_2 and CO were both detected on the surface in the 1990s as well.

Although N_2 and CO have similar volatilities, thereby making them competitive for the major species in Pluto's atmosphere, the existing constraints on CO abundance both in the surface ices and in the atmosphere agree that N_2 must be the dominant atmospheric species. Pluto's surface temperature has been measured by various infrared and radio astronomy techniques – apparently, the N_2 ice regions are presently near 40 K, and the less volatile, CH_4 ice/"dirt" regions are closer to 60 K. The fact that the N_2 ice is dominant on the surface and sublimating at a temperature near 40 K tells us that the surface pressure regime inferred from the stellar occultation is in agreement what one would predict from nearly pure N_2-ice patches, but far lower than we had expected for any of our 1980–1981 era model atmospheres. If the atmospheric collapse model is correct, the surface temperature at aphelion could drop as low as 25–30 K.

Larry Trafton and I went on to do a good deal of other Pluto work in the 1980s and 1990s, much of it together, and also a few projects individually. Together we published studies of Pluto's atmospheric dynamics, its ultraviolet spectrum, and (in conjunction with Canadian ex-pat Randy Gladstone), a refinement of the atmospheric collapse model that predicted the collapse would occur no earlier than the late 1990s, and possibly as late as the 2020s. On his own, Larry extended the global collapse work to Triton, Neptune's largest moon, which has an atmosphere strikingly similar to Pluto's. Larry also undertook detailed studies of Pluto's atmospheric escape rate, and the microphysics of the surface–atmosphere interaction. I branched out into other Pluto work, including searches for additional satellites, studies of the rate at which Pluto should be cratered by comets and Kuiper Belt Objects, studies of the origin of Pluto and the general problem of accretion in the Kuiper Belt, and the issue of whether Pluto's interior should be internally homogeneous or differentiated (I argued for differentiation).

Fall 2001: a choice of futures

The degree to which Pluto's atmosphere will collapse is not yet certain. Indeed, John Stansberry and Roger Yelle, two talented planetary scientists at the University of Arizona in Tucson, developed an alternative to atmospheric collapse in the late 1990s. Their theory goes as follows: The N_2 ice that covers much of Pluto's surface is known to come in two different crystal types, called alpha-nitrogen and beta-nitrogen. The crystal form that the N_2 actually occurs in depends on the temperature of the ice – above 35.6 K it will be beta-nitrogen, below 35.6 K it will be alpha-nitrogen. The conversion from beta to alpha as Pluto recedes from the Sun and cools releases energy (in reordering the crystal lattice). According to the Stansberry–Yelle model, the emissive properties of alpha- and beta-nitrogen, which play a major role in setting the N_2 ice temperature, are such that the N_2 ice temperature (like a thermostat) never varies more than a few degrees from 35.6 K. If this model is right, then the atmosphere will not fully collapse, because the temperature (and therefore the atmospheric pressure and abundance) will never fall very far, even when Pluto is at its most distant point from the Sun.

I find this model very clever, in the best sense of the word. However, I do not believe it is likely to be right for a number of reasons. The primary objection I have is that the model relies on some simplifying assumptions that are unlikely to be realistic on the real-world Pluto (for example, the model ignores the effect of CH_4 and CO frosts interacting with the N_2). I also like to point out that Pluto's surface albedo is known to have grown steadily darker as it approached the Sun in the 1950s, 1960s, and 1970s, as if the recently snowed-out atmosphere of the last orbital cycle was re-sublimating into the atmosphere as the planet warmed, which would not have occurred in the Stansberry–Yelle scenario.

Resolution

What will actually happen to Pluto's atmosphere? Will it collapse dramatically? Time, fortunately, *will* tell.

From Earth we can watch for a brightening in Pluto's reflectivity, which would also likely be accompanied by a reduction in its redness, both caused by the deposition of frost if the atmosphere begins to snow out. (Current estimates of the atmospheric pressure correspond to an equivalent snow-out depth of 0.05 to 1 centimeters, what we would call a light frost in Colorado.)

Because such a thin frost might be hard to detect, we might instead have to rely on chance stellar occultations, which are rare. There was one in 2002, and statistics tell us to expect additional events about every five to ten years. The signature of atmospheric collapse in an occultation dataset would be a much sharper reduction in the starlight as the occultation begins and ends. Why? With less air the role of refraction in making the occultation signal decline gently would be much diminished.

Another way to determine whether the atmospheric collapse theory is right or wrong is to go to Pluto, as NASA is now planning to do. After more than a decade of recommendations by the scientific community and study by NASA, a project and team called *New Horizons* (which I have the privilege of leading) has been funded to start development of a Pluto mission that would launch in early 2006 and arrive between 2015 and 2017 (see: http://pluto.jhuapl.edu) (Even traveling at 100 times the speed of a jetliner, as *New Horizons* will, five billion kilometers is a very long haul.)

When we arrive at Pluto, the cameras, spectrometers, and other instruments aboard *New Horizons* will measure the atmospheric pressure directly to see if the collapse has occurred. Even if the collapse has not occurred, we can search for evidence of past collapse cycles by looking for Plutonian analogs of the polar layered terrain that records such cycles on Mars. If the collapse has occurred, we will know it from our atmospheric measurements and surface reflectivity studies.

In any event, someday in the not too distant future, we *will* know whether Pluto's atmosphere waxes and wanes with its orbital cycle, or whether the finely tuned Stansberry–Yelle thermostat instead regulates the pressure, keeping it fixed near its present value.

This is how science works. Theories are tested, both by observation and by constructive critique. Refinements are tested the same way. Only the correct model can survive the crucible of such tests over time. This wonderful feedback loop is the key to science's success in explaining the natural world, and it is the method by which we shall someday learn the fate of Pluto's atmosphere.

Science, self-testing, self-correcting, is a wonderful invention of the human species. And it is a marvelous adventure illuminating the world that surrounds us.

Further reading

Binzel, R. P. (1990). Pluto, *Scientific American*, **262**: 50–58.

Stern, S. A. and J. Mitton (1999). *Pluto & Charon: Ice Worlds on the Ragged Edge of the Solar System*, New York: John Wiley, 216 pp.

Stern, S. A. and L. M. Trafton (1984). Constraints on Bulk composition, Seasonal Variation, and Global Dynamics of Pluto's Atmosphere, *Icarus*, **57**: 231–240.

Closely Watched Trains
Bohumil ~~Hr~~ Hrabal